The Springer History of Mathematics Education Series aims to make available to scholars and to teachers at all levels around the world the fruits of outstanding research into the history of mathematics education; to provide historical syntheses of comparative research on important themes in mathematics education; and to establish greater interest in the history of mathematics education.

History of Mathematics Education

Series Editors

Nerida F. Ellerton, Mathematics Department, Illinois State University, Normal, US

M. A. Ken Clements, Mathematics Department, Illinois State University, Norm
USA

Marc Moyon • Dominique Tournès •
Snezana Lawrence
Editors

Teaching Mathematics Through Historically-Based Activities

Experiments in French Classrooms

Springer

Editors

Marc Moyon
Institut XLIM UMR CNRS 7252
IREM de Limoges
Université de Limoges
Limoges, France

Dominique Tournès
Parc Technologique Universitaire, LIM
IREMI de la Réunion
Université de la Réunion
Sainte-Clotilde, Réunion

Snezana Lawrence
Independent Scholar
Cambridgeshire, UK

ISSN 2509-9736 ISSN 2509-9744 (electronic)
History of Mathematics Education
ISBN 978-3-032-16989-1 ISBN 978-3-032-16990-7 (eBook)
https://doi.org/10.1007/978-3-032-16990-7

Translation from the French language edition: "Passerelles: enseigner les mathématiques par leur histoire au cycle 3" by Marc Moyon et al., © ARPEME-APMEP 2019. Published by ARPEME. All Rights Reserved.

This Springer imprint is published by the registered company Springer Nature Switzerland AG
The registered company address is: Gewerbestrasse 11, 6330 Cham, Switzerland

Preface to the Series

Books in Springer's Series on the history of mathematics education comprise scholarly works on a wide variety of themes, prepared by authors from around the world. We expect that authors contributing to books in this Springer Series will go beyond top-down approaches to history, so that emphasis will be placed on the learning, teaching, assessment, and wider cultural and societal issues associated with mathematics education at all levels.

In addition to generating texts on the history of mathematics education written by authors in various nations, an important aim of the series is to develop and report syntheses of historical research that have already been carried out in different parts of the world with respect to important themes in mathematics education—like, for example, "Historical Perspectives on how Language Factors Influence Mathematics Teaching and Learning," and "Historically Important Theories which have Influenced the Learning and Teaching of Mathematics."

The mission for the series can be summarized as:

- To make available to scholars and to teachers at all levels around the world the fruits of outstanding research into the history of mathematics education
- To provide historical syntheses of comparative research on important themes in mathematics education
- To establish greater interest in the history of mathematics education

The authors of each chapter in this book describe how young children in schools across France were introduced to different mathematical concepts through the lens of historical documents and traditions. Teachers aimed to help learners mathematize the historical concepts being introduced. Readers will be amazed not only at what the children were asked to learn but also at the teaching methods used. Throughout the book, details of how teachers cooperated with each other and with their students are provided, with the teachers aiming to ensure that the links with history have been faithfully maintained and documented. Each chapter should challenge all mathematics educators to reflect on the rich resources embraced by the history of mathematics.

We hope that the series will continue to provide a multi-layered canvas portraying important details of mathematics education from the past, while at the same time presenting historical insights that can support the future. This is a canvas which can never be complete, for today's mathematics education becomes history for tomorrow. A single snapshot of mathematics education today is, by contrast with this canvas, flat and unidimensional—a mere pixel in a detailed image. We encourage readers both to explore and to contribute to the detailed image which is beginning to take shape on the canvas for this series.

Any scholar contemplating the preparation of a book for the series is invited to contact Springer Nature directly.

Normal, IL, USA
December 2025

Nerida F. Ellerton
M. A. (Ken) Clements

Introduction

This book is authored by a group of mathematics teachers and educators. It serves as an introduction to others who are interested in the history of mathematics. The book shows in some detail the ways the history of mathematics was integrated into mathematics teaching, with detailed reflections on the process, stating the routes to explore further. The book is based on nine experiments—unlike the *Nine Chapters of Mathematical Art* (Chemla & Guo, 2004), it does not state new mathematics—but shows how old can inspire learning in the discipline today.

The use of history in the teaching of mathematics was initiated for many reasons, but perhaps mainly to inspire and motivate pupils to see mathematics as a discipline that is constantly growing and, whilst universally recognized for its cogency, is also dependent on context in which it was developed. These two key benefits, often highlighted, are: the increased motivation it fosters in pupils through the way it enables them to explore mathematical ideas through engaging historical narratives, and the possibility of developing a learning landscape. In the case of the latter, both pupils and teachers are allowed to roam freely in order to find those episodes that would further take them to make connections within mathematics to always new routes, thus prolonging desire to learn, and leading to a potentially never ending journey. However, such a summary may be far too limited! And no number of articles that discuss the aims, objectives, and outcomes of using history in the teaching of mathematics can help a practitioner try doing this in their classroom themselves. So, to uncover a broader range of reasons, as well as see what can be done in an actual mathematics classroom, we have put together this book as a collection of the already mentioned nine experiments. These were conducted in French classrooms, and within the following pages, you will find both the resources used or a mention where and how you can start making your own to try similar things. You will also find reflections on how effective such approaches were for the French colleagues.

Before diving into the details, it is worth pausing and thinking about a broader context. We have been asked to think, by the reviewers of our book, about a seeming contradiction. The question they posed is this: How come the international assessments such as TIMSS and PISA consistently report that French students perform relatively poorly compared to their peers worldwide, yet, the classroom episodes presented in the following chapters, from pupils aged 8 to 12, seem to show strikingly high levels of mathematical thinking and understanding? This is a good question. One answer may be that experiments we note here are small in number and do not make enough of impact on a large country such as France. But also we could speculate that in classrooms where mathematics is taught in close connection with its history, students attain deeper engagement and insight and hence are able to achieve better results in mathematics. These seem to be fairly open questions, and answering them would require a national, longitudinal study. For now, we turn to the evidence at hand: material drawn from nine French schools that invites us to reflect more deeply on what it means to "do well" in mathematics.

The Background: History of Mathematics and Mathematics Education

Over the past half-century (since 1970s), the history of mathematics in mathematics education has grown into an entire field of study. This small but nicely defined field lies somewhere between the history of mathematics itself and mathematics education; it is a fertile and interesting place to explore! I can speak about this firsthand, as I served as the Chair of a prominent international study group called *History and Pedagogy of Mathematics* (HPM for short). This group was founded in 1976 and has been affiliated with the *International Commission on Mathematics Instruction* (ICMI). The ICMI holds congresses every four years, where the HPM is represented as an affiliated organization. The ICMI itself is part of the *International Mathematics Union* (IMU), the singular global platform and network of mathematicians who branch into all different directions in their exploration of mathematics.

Research into the ways the history of mathematics can be applied in the classroom reveals numerous approaches to achieve greater engagement in mathematics (Chorlay et al., 2022; Clark et al., 2019). Additional benefits for both pupils and teachers are the broader view of the discipline, and a much greater refinement in understanding mathematical concepts. None of these are easy to achieve, and so various networks have sprung up internationally by teachers and scholars in search of communication and peer support on their ways to developing these ideas and methods further. However, the number one problem teachers and teacher trainers cite as a reason for not trying the history in mathematics classroom is time management. Both time spent in a classroom and the time out it are very limited to develop knowledge as well as skill to use history effectively. The time in the classroom is usually never sufficient to cover all a teacher wants to cover; the time out of the classroom is also short when marking and preparation are taken into account, so learning about the history of mathematical sciences can seem like a time not too well spent (Siu, 2006; Moyon, 2022).

Leading the way in this field where such support and cooperative projects are possible can be found in the French tradition of IREMs (*Instituts de Recherche sur l'Enseignement des Mathématiques*—Mathematics Education Research Institutes). These are regional research networks dedicated to all aspects of research into mathematics education and were founded in 1969. These networks usually have a centre in a university, but teachers from regional schools participate and work with researchers from universities. History of mathematics is one of the sections one can join in these networks.

The nine chapters presented here all come from teachers and academics belonging to this network of IREMs. Here they generously share the rich experiences they had in their classrooms, with the wider audiences. The showcased experiments were done over a period of several years and explore very different mathematical topics, objects, and traditions that shaped mathematics and mathematical practices throughout history, and in different geographical areas. This is, in other words, an example of a mathematical curriculum rooted in history. The experiments in this book demonstrate how historical context can add depth and excitement to both teaching and learning of mathematics. Across the nine chapters, we encounter the ancient Mayan mathematics, the mechanics of abacus, mechanization of calculus, ratios and fractions, Plato's ideas related to mathematics and its learning, the innovations of the Revolutionary French Calendar, surveying techniques, Leonardo's mathematical opus, and geometry of fortifications. Each example is presented in a clear and structured format, detailing the history, then the experiments undertaken with the challenges they posed, and opportunities they offered for learning mathematics in a history-rich context. There is also a narrative of a clear and practical guidance here for those teachers and teacher educators that wish to try this type of approach themselves, and ample guidance on how to adopt and apply historical approach in their own classrooms.

Summary of Chapters

In Chap. 1, Mayan mathematics is examined and how it could be applied in a classroom. Alice Morales, Jerome Capitan, Ludovic Jollet, Anne Jorioz, Anne-Marie Marmier, and Jean-Baptiste Meilhan are members of the *History in the Teaching of Mathematics* research team at IREMI of Grenoble. Their work highlights the educational value of exploring numerical systems from different cultures, with a focus on the ancient Maya's vigesimal (base-20) system. They see this system as a fertile ground from which to deepen pupils' understanding of the decimal system. By engaging with alternative numerical bases, pupils not only enhance their mathematical skills but also gain insights into the cultural and historical contexts that shaped these systems.

The Maya civilization, which flourished in Central America between 300 and 900 AD, achieved remarkable advances in architecture, astronomy, and agriculture before its decline due to environmental challenges and the Spanish conquest. Their numeration system, which incorporated a placeholder for zero and operated in base-20, offers unique opportunities for teaching. Through hands-on activities, pupils explored Mayan mathematics, discovering how the system worked while grappling with the differences between it and the familiar base-10 framework. This often led to insightful moments as pupils mixed the two systems, illuminating their own assumptions about numbers.

Additionally, and perhaps surprisingly, activities incorporating the Yucatec language helped pupils understand the structure of Mayan number construction, emphasizing the importance of powers of a base. This immersive approach not only boosted engagement but also demonstrated how integrating historical and cultural narratives into math education can deepen understanding, foster curiosity, and broaden perspectives.

Chapter 2 was written by Nathalie Daval and Dominique Tournès, working within the IREM of La Réunion with other teachers as listed in the chapter introduction. Here we learn about how to teach mathematics through dealing with mathematical tools. Tools that could help calculations or procedures have always been important for mathematical practitioners. This begs a question: "What will happen with mathematics and artificial intelligence in the future?" To consider such questions, and possible answers, it may be a good idea to consider what happened with the introduction of new such tools by practitioners from the past.

This chapter highlights the importance of tangible mathematical tools, such as abaci in teaching decimal numeration and basic calculation techniques. We see how important it is for children to first manipulate physical objects, like tokens or pebbles, before moving on to abstract symbols. The focus of this chapter is on token abaci, which have been used in the West for over 2,000 years, distinguishing them from the bead abaci common in the East. Unlike bead abaci, token ones allow unlimited tokens in each row or column, offering unique flexibility.

Archaeological evidence reveals about 30 abacus fragments from the Greek world, primarily associated with money changers and merchants. The oldest known example is a white marble slab from the fifth or fourth century BC, engraved with symbols and designed around a base-10 system with a sub-base of 5, inspired by human finger counting. This system also uses subitizing—the ability to instantly recognize small quantities—typically limited to groups of four or fewer.

The token abacus was later adopted by the Etruscans and Romans, as seen in artefacts like an Etruscan calculator and Roman bas-reliefs depicting calculators in elite households. Even after the introduction of Indian calculation methods in the tenth century, widespread illiteracy and the impracticality of Roman numerals ensured the abacus remained in use. A notable example is Gerbert of Aurillac's (c.946–1003) innovation of *Gerbert's abacus*, which demonstrated the tool's enduring relevance. Gerbert was, apart from being a mathematician, also a Pope (known as Sylvester II), and popularized Indo-Arabic numerals.

Chapter 3, written by Frédérique Plantevin with the help of the members of a group of researchers from the IREM of Brest, examines the role of mechanical instruments in teaching arithmetic operations—addition, subtraction, multiplication, and division—to pupils aged 9–11, while also appealing to a wider audience.

It begins with a historical overview of mechanization in arithmetic, initiated by Wilhelm Schickard (1592–1635) and Blaise Pascal (1623–1662) in the seventeenth century, who both created first mechanical machines for automatic addition. Work presented here shows how pupils can explore mathematical concepts by using different machines (appropriate to the century in which they were invented). Pupils can then also create user manuals to develop algorithmic thinking and engage in discussions about the differences between different ways of using calculating machines. This chapter underscores the significance of historical instruments in enriching mathematical understanding, promoting an interdisciplinary approach that fosters practical intelligence, and makes learning mathematics engaging.

In Chap. 5, Sylviane Schwer and others from IREM of Paris Nord present us with interesting work on the number ratios and the approaches to teaching fractions. Until the nineteenth century, arithmetic education in Europe primarily focused on basic operations using abaci with tokens for young monks and merchants. The revival of Boethius' (c.480–524) *Arithmetic* (first printed in 1488) and *Music* (edited first in 1491 or 1492) rekindled interest in Pythagorean numerical concepts, emphasizing the properties and relationships of numbers, significantly influencing medieval mathematical education. The chapter traces the evolution of natural numbers and fractions through Pythagorean thought, laying a foundation for arithmetic teaching that persisted into the seventeenth and eighteenth centuries. By using simple token manipulations, pupils can grasp fractions without feeling overwhelmed, promoting critical thinking and problem-solving skills.

The roles of mathematicians have historically intertwined with philosophy, physics, and engineering. Two contrasting Greek perspectives emerged: Parmenides (date unknown–460BC), who advocated for an unchanging reality, and Heraclitus (c.600–c.500BC), who asserted perpetual change. Pythagoras (c.570–c.495BC) emphasized numbers as essential to understanding the universe, influencing later thinkers like Aristotle (382–322BC). Despite facing challenges, Pythagorean thought was preserved by Nicomachus (c.60–c.120AD) after it was revived by Plato (428/427–348BC) and Aristotle. During the Middle Ages, the Church structured university knowledge around the seven liberal arts—grammar, rhetoric, logic, geometry, arithmetic, music, and astronomy. Four latter were often considered to be mathematical (Dee, 1570). By the seventeenth century, definitions of numbers evolved, culminating in a structured understanding of various types, affirming that humanity indeed arithmetizes and has been doing it since ancient times. Additionally, in this chapter, the teaching strategies for ratios are proposed, using interactive activities to enhance understanding and encourage collaboration between mathematics and computer science education.

Chapter 4, by a group M.A.T.H. of the Paris' IREM, led by Renaud Chorlay, presents a three-session educational sequence for fifth and sixth graders based on a classic mathematical problem from Plato's *Meno*. In the dialogue, Socrates challenges a slave to create a square with double the area of an initial square, but the slave mistakenly suggests doubling the side length, resulting in a square with quadruple the area. Socrates demonstrates the correct method, promoting discussions on geometric vocabulary and area comparison. The approach aims to enhance pupils' argumentative skills in geometry while integrating language comprehension for interdisciplinary learning.

The historical context primarily aids teachers and does not directly feature in pupil sessions. During Plato's era, and mathematical texts like Euclid's *Elements* that emerged later, inquiries often took a backseat to philosophical themes. The chapter emphasizes a socio-constructivist approach, positioning pupils as experts to evaluate existing knowledge and critique formulations while confronting epistemological obstacles, such as distinguishing between contour lengths and areas.

The sequence incorporates active reading strategies and guided participation. While the first session posed challenges due to text obscurity and ambiguous terminology, subsequent sessions allowed for more structured engagement with mathematical problems. Despite varied responses, the use of numerical frameworks was surprisingly limited.

Chapter 6 is written by Frédéric de Ligt and his colleagues from the IREM of Poitiers. They explore the significant challenges young pupils face with the decimal system, fractions, and conversions in mathematics, contextualized through a historical episode from the French Revolution. It discusses the transition from royal to universal measurement standards following the abolition of privileges in 1789, encapsulated in the phrase "One king, one law; one weight and one measure". The initiative from May 8, 1790, led to the formation of a commission to unify the measurement system, establishing the metre as a fundamental unit, with contributions from notable figures like Condorcet (1743–1794) and Lavoisier (1743–1794). The chapter highlights the historical roots of the decimal system, dating back to Simon Stevin (1548–1620) in the sixteenth century, emphasizing its relevance in everyday life.

It also details the Calendar Commission which aimed to reform the calendar to standardize timekeeping. The year was defined as 365 days, 5 hours, 48 minutes, and 49 seconds, divided into 12 months and three decades of 10 days, with additional "sans-culottides" days. These were the five complementary days in a common year as the months of the Republican Calendar were 30-day long (rather than as in the usual calendar alternating from 30 to 31). Despite initial resistance, modifications were made to ease almanac creation, creatively naming months after agricultural themes and introducing a decimal time division. However, practical challenges hindered widespread acceptance. The chapter encourages pupils to appreciate the connection between mathematics, history, and scientific progress.

Chapter 7 by Marc Troudet, Cécile Chovet, Saida Boué, and Pauline Mirmand, who work within IREMI of Grenoble and IREM of Lyon worked on introducing their pupils to concepts relating to practical mathematics and surveying. To effectively grasp areas, pupils should practice measuring regular surfaces like doors and rooms, which helps them become familiar with the metric system—an essential foundation for future studies. Surveying was defined in 1852 as measuring land relative to a unit, highlighting practical geometric applications. After a decline in concrete problem-solving during the 1960s and 70s, contemporary teaching methods have revived this approach, addressing common pupil difficulties through practical problem-solving and progressive learning over three years.

The teaching of surveying in primary schools evolved from the Ancien Régime, gaining official recognition in 1816 for qualified teachers. By 1833, practical geometry was integrated into primary education as vital for industrial professions. Some educators emphasized hands-on experiences, using community resources for measuring activities. Fieldwork became mandatory in 1893, and teacher training prioritized practical techniques.

Observations over two years of instruction reveal that maintaining the same problem across years motivates pupils while accommodating varying learning levels. This hands-on approach, utilizing software and instruments, enhances engagement and fosters peer interaction. Overall, surveying supports mathematical understanding and facilitates interdisciplinary connections between primary and secondary education, reinforcing the importance of practical skills in learning.

The penultimate, Chap. 8, written by Marc Moyon, working with a group of teachers from the IREM of Limoges, focuses on using Leonardo da Vinci's (1452–1519) mathematical work to teach geometric concepts such as intersection points, parallelism, and perpendicularity. These were presented through activities for pupils aged 8–11. Lessons conducted and analysed involve reproducing complex figures with rulers and compasses, blending geometric learning with historical and scientific contexts. Leonardo, born in Vinci, Tuscany, was self-taught, gaining knowledge through his personal library and interactions with scholars. After becoming an apprentice to Andrea Verrocchio (c.1435–1488) in Florence, he joined the painters' guild and later worked in Milan under Duke Ludovico Sforza (1452–1508), where his interest in mathematics grew.

Leonardo's collaboration with mathematician Fra Luca Pacioli (c.1447–1517) in 1496 led to deep explorations of Euclid's *Elements* after which Leonardo created beautiful and complex images including the famous *Vitruvian Man* (c.1490) which shows ideal proportions of a human body. Leonardo's work emphasized ratios, proportions, and area transformations, demonstrated through methods accessible to pupils, such as sliding figures and cutting and pasting geometric shapes to compare areas.

The chapter draws from folio 471verso of Leonardo's *Codex Atlanticus* (1478–1519), a compilation of his notes and sketches, showcasing his lifelong curiosity about geometry. Leonardo's interdisciplinary work connects subjects like mathematics, history, and geography, making him an ideal figure for exploring the broader historical context and advancing pupils' understanding of how past processes shape modern society.

Finally, Chap. 9 by Frédéric Métin et al., from IREM of Dijon, emphasizes the interdependence of theory and practice in learning, particularly through the lens of seventeenth-century military architecture, which offers a wealth of geometric insights. Citadel designs, notably "Vauban stars", models of fortification due to Sébastien Le Prestre, *seigneur de Vauban* (1633–1707) provide complex yet regular forms suitable for visual analysis via plans and aerial images. Part of this project was an exploration of scientific applications in military contexts, where pupils learnt how to connect geometry as a universal language crucial for military strategy.

The seventeenth-century evolution of fortifications showcases the ongoing relationship between attack and defence, as engineers increasingly utilized Euclidean geometry to validate their designs. One of the episodes referred to in the project was investigation of a critical transition in development of artillery, where advancements in cast iron cannons transformed siege warfare, making traditional towers vulnerable to precise fire. This prompted the creation of pentagonal bastions in Italy, designed to mitigate blind spots and incorporate geometric principles, though practical applications.

But history offers some interesting and more informed visions of the present as well as the possible futures. Project supported thinking about these too-modern tools like OpenStreetMap promote the practical application of geometry in historical contexts, demonstrating its relevance across disciplines.

Conclusion

A very diverse collection of topics, with the range of pupils that were involved in these projects, should make every teacher more equipped to start the journey of using the history of mathematics in their own teaching. These were very practical and very precisely defined projects—and teachers who consider them and perhaps try something similar will also have some benefit of learning in advance, from the reflections of the authors of these chapters, where the pitfalls and possible problems may lie.

The hope is that you will look at this book, pick perhaps one chapter, and try things described there; try another, and perhaps a third, and eventually design your own project and find colleagues you could work with on it. And we will then look forward to meeting you at some of the networks mentioned at the beginning of this chapter.

Cambridgeshire, UK Snezana Lawrence
December 2024

References

Chemla, K., & Guo, S. (2004). *Les neuf chapitres: le classique mathématique de la Chine ancienne et ses commentaires*. Dunod.

Chorlay, R., Clark, K., & Tzanakis, C. (2022). History of mathematics in mathematics education: Recent developments in the field. *ZDM – Mathematics Education*, *54*(7), 1407–1420.

Clark, K. M., Kjeldsen, T. H., Schorcht, S., & Tzanakis, C. (2019). History of mathematics in mathematics education – An overview. *Mathematica Didactica*, *42*(1), 1–26.

Dee, J. (1570). *The Mathematicall Praeface to Elements of Geometrie of Euclid of Megara*. John Daye.

HPM. (no date). https://hpm.sites.uu.nl

ICMI. (no date). https://www.mathunion.org/icmi

IMU. (no date). https://www.mathunion.org/

IREM. (no date). https://www.univ-irem.fr

Moyon, M. (2022). Desire of teachers and realities in textbooks: Dealing with history of mathematics in the new French curriculum and its impact on teacher training. *ZDM – Mathematics Education*, *54*(7), 1613–1630.

Siu, M. (2006). No, I don't use history of mathematics in my class: Why? In F. Furinghetti, S. Kaijser, & C. Tzanakis, (Eds.), *Proceedings HPM 2004 & ESU4 (revised edition)* (pp. 268–277). University of Crete.

General Organization of the Book

Each chapter of this book follows the same pattern, so that readers can easily find their way around:

1. In the "Historical Context" section, the authors aim to clearly present the essential background of the mathematical episode, providing just enough information to understand the key issues of the period. They focus on the sociopolitical context, previous advancements, key figures, and major discoveries, helping the reader grasp the relevance of the topic.
2. In the "Scientific and Pedagogical Aspects" section, the authors aim to highlight the most significant elements behind their choice of activity, particularly by showing how didactic and historical studies can complement each other.
3. The "Implementation in the Classroom" refers to the practical application of teaching approaches, strategies, and activities designed, with history of mathematics, to engage pupils and effectively convey the subject matter in an interactive learning environment. These experiments are all carried out in classes from the "consolidation cycle" (end of primary school and beginning of middle school), with pupils aged between 8 and 11 years old.
4. In the "Ways to Go Further" section, the authors suggest several ideas to deepen the reflection and implement additional pedagogical strategies, either by increasing the difficulty or adapting them for different audiences.
5. The "Acknowledgements" section outlines the writing conditions of each chapter and acknowledges all those involved in the educational project, each according to their role.

Each chapter ends with its own appendix, often providing the material needed to implement the proposed teaching session(s), and references. Everything is designed to facilitate reading and the provision of documents for use in the classroom or in teacher training. All these documents are available (in French but some of them are in English) thanks to the French IREM network website, along with bibliographical references freely available on the web https://www.univ-irem.fr/-passerelles-enseigner-les-mathematiques-par-leur-histoire-en-cycle-3- (accessed on June 30, 2025).

This book contains numerous quotations, especially originally in French. Unless otherwise indicated, all translations into English are the authors' own.

It's also worth noting that the first appendix of each chapter sets out the chapter's knowledge and know-hows, as well as the skills specifically addressed in that chapter. These skills follow the six general skills put forward by the French Ministry of Education, namely:

- Search and research skills
- Reasoning
- Computational skills
- Mathematical communication skills
- Mathematical representation ability
- Modeling

The following timeline (Fig. 1) includes each chapter in the appropriate historical period. It could be easier to understand the historical context of each chapter.

Finally, editors would like to express their sincere gratitude to the Inter-IREM Commission on "Epistemology and History of Mathematics," to the broader IREM network (Institutes for Research on Mathematics Education), to ARPEME (Association for the Development and Dissemination of Educational Resources for Teaching Mathematics in Schools) that published the French version of this book, to ADERHEM (Association for the Development of Studies and Research in the History and Epistemology of Mathematics), and to the APMEP (Association of Mathematics Teachers in Public Education) for their invaluable support of the "Passerelles" project. This initiative aimed to teach mathematics through its history, offering children aged 8–11 the opportunity to explore mathematical concepts within their historical context.

We would also like to especially thank the Academy of Sciences (Paris, France) for awarding the project its "Scientific Teaching Book Prize" in 2019, a recognition that has been instrumental in furthering the impact of this educational journey. We are deeply grateful to all the contributors, each in their own capacity, whose dedication and commitment have been crucial to the success of this project. Their collective efforts have been essential in making this initiative possible across nine regions in France.

Marc Moyon
Dominique Tournès
Snezana Lawrence

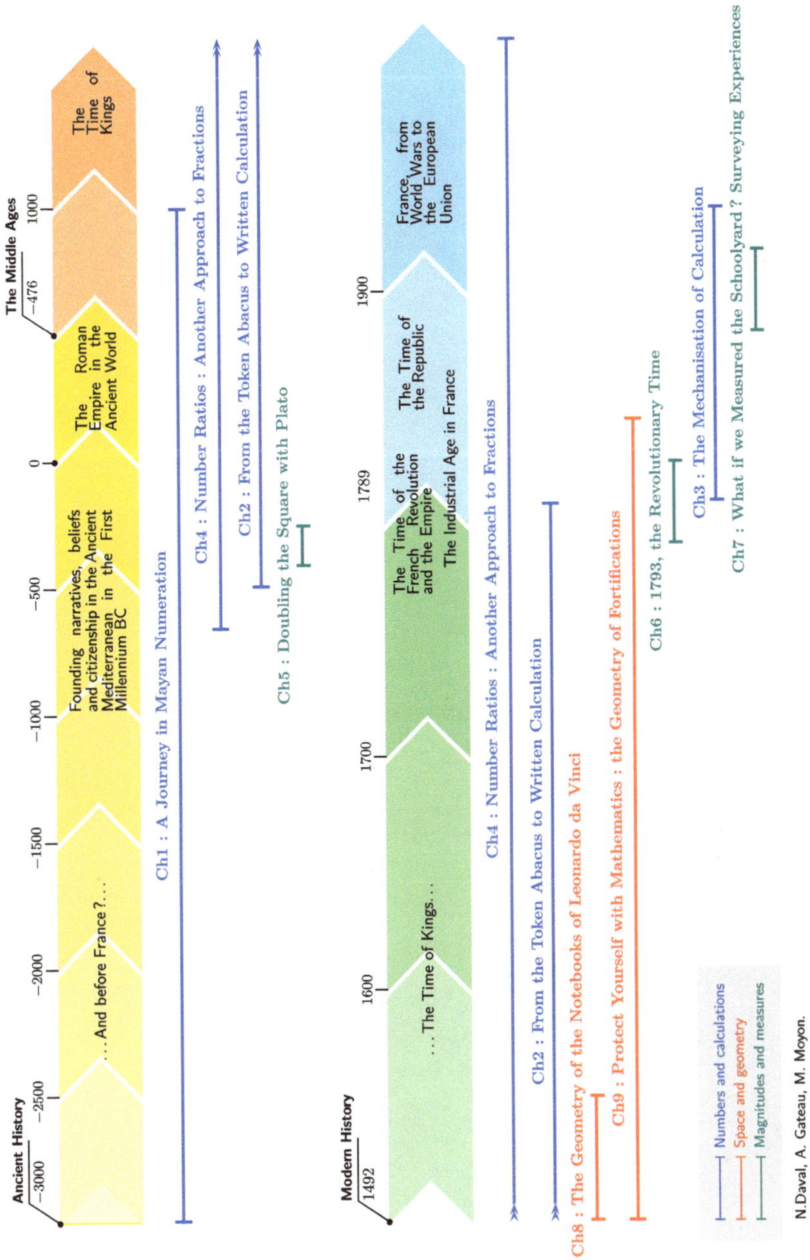

Fig. 1 Timeline describing the whole book by part and chapter

Contents

Part I Numbers and Calculations

Part II Magnitudes and Measures

Editors and Contributors

About the Editors

Marc Moyon is Professor of History of Mathematics at the University of Limoges and a member of the XLIM Institute (UMR 7252, CNRS). He is also *Professor Catedrático Convidado* at the University of Aveiro (Portugal) within CIDMA (Center for Research and Development in Mathematics and Applications). His research focuses on medieval mathematics and the intercultural transmission of scientific knowledge, especially through Arabic-Latin translations, as well as the history of mathematics education in the nineteenth and twentieth centuries. He actively contributes to introducing historical perspectives into mathematics education and teachers training, notably within the national French IREM network (Institute of Research on Mathematics Education), and the International Study Group on the Relations between the History and Pedagogy of Mathematics (HPM) affiliated to ICMI.

Dominique Tournès is professor of mathematics and history of mathematics at the University of Reunion Island (France), in the Computer Science and Mathematics Laboratory. As a historian of mathematics, his research focuses on the calculation methods and instruments that preceded the computer in the period 1750–1950. For many years, he has been involved in using history to improve the training of mathematics teachers.

Snezana Lawrence is an independent scholar, a mathematical historian, and was a teacher, teacher educator, and taught mathematics in schools and universities in England until recently. She specializes in mathematics of the revolutionary period in France, and more recently mathematics before and during the Cold War. Snezana was the Chair of the HPM (History and Pedagogy of Mathematics International Study Group, 2020–2024); first Education Officer of the BSHM (British Society for the History of Mathematics); and the Assistant Editor of the *British Journal for the History of Mathematics* (Taylor & Francis). She is particularly skilful in describing mathematical concepts in a captivating and easy-to-follow narrative. Her latest book is *A Little History of Mathematics* (Yale University Press, 2025).

Contributors

Dominique Bernard École Jacques Prévert, Brest, France

Sandrine Bonnet-Ventura Groupe scolaire des Grandes Varennes, rue Gabriel Charlopeau, La Rochelle, France

Saida Boué École primaire publique, Saint Cyr sur le Rhône, France

Jerome Capitan IREMI de Grenoble, Saint Martin d'Hères, France

Grégoire Charles École Ormeteau, Aulnay-sous-Bois, France

Renaud Chorlay IREMS de Paris, Université Paris Cité, Paris, France

Nathalie Daval Parc Technologique Universitaire, LIM, IREMI de la Réunion, Université de la Réunion, Sainte-Clotilde, Réunion

Jérome Dufour (†) Collège Cabanis, Brive-la-Gaillarde, France

Catherine Dugain-Abrousse École élémentaire Raphaël Barquisseau, Saint-Louis, France

Deborah Fau École Ormeteau, Aulnay-sous-Bois, France

Audrey Fayollat École Jacques-Yves Cousteau, Vouneuil-sous-Biard, France

Chantal Fourest Cité scolaire d'Arsonval, Brive-la-Gaillarde, France

Véronique Fustec Collège Saint-Pol-Roux, Brest, France

Matthieu Gaud Collège Pierre Mendès France, La Rochelle, France

Alexis Gautreau Cité scolaire Rodin, Paris, France

Tiffany Gérin École Ormeteau, Aulnay-sous-Bois, France

Murielle Geslin École du Champ-de-Foire, Plougastel Daoulas, France

Amandine Grondin École élémentaire publique Léonide Letoullec, Le Port, France

Sylvie Grondin École primaire de Piton A, Saint-Leu, France

Yan Guede École Ormeteau, Aulnay-sous-Bois, France

Marie-Jeanne Guillotin École élémentaire publique Martin Luther King, Saint-Pierre, France

Dominique Héguiaphal École élémentaire publique d'application Arago, Paris, France

Ludovic Jollet IREMI de Grenoble, Saint Martin d'Hères, France

Anne Jorioz IREMI de Grenoble, Saint Martin d'Hères, France

Snezana Lawrence Independent Scholar, Cambridgeshire, UK

Yohan Lefèbvre Collège Saint-Pol-Roux, Brest, France

Véronique Lefranc École Roger-Gouffault, Brive-la-Gaillarde, France

Marc Le Pors Circonscription de Brest Ville, Brest, France

Yann Le Thénaff Collège Saint-Pol-Roux, Brest, France

Catherine Levy École Ormeteau, Aulnay-sous-Bois, France

Frédéric de Ligt IREM&S de Poitiers, Poitiers, France

Anne-Marie Marmier IREMI de Grenoble, Saint Martin d'Hères, France

Jean-Baptiste Meilhan IREMI de Grenoble, Université Grenoble Alpes, Grenoble, France

Frédéric Métin IREM de Dijon, Université Bourgogne-Europe, Dijon, France

Pauline Mirmand École Édouard Herriot Prainet 1, Décines-Charpieu, France

Alice Morales IREMI de Grenoble, Université Grenoble Alpes, Grenoble, France

Marc Moyon Institut XLIM UMR CNRS 7252, IREM de Limoges, Université de Limoges, Limoges, France

Frédérique Plantevin IREM de Brest, Université de Bretagne Occidentale, Brest, France

Cyril Redondo Collège Pierre Mendès France, La Rochelle, France

Fabienne Romian Collège Gérard-Philippe, Aulnay-sous-Bois, France

Valérie Rosier-David Cité scolaire d'Arsonval, Brive-la-Gaillarde, France

Sylviane R. Schwer IREM Paris Nord, Université Sorbonne Paris Nord, Villetaneuse, France

David Somdecoste École Louis Pons, Brive-la-Gaillarde, France

Luc Tiennot IREMI de la Réunion, Parc technologique universitaire, Sainte-Clotilde, France

Dominique Tournès Parc Technologique Universitaire, LIM, IREMI de la Réunion, Université de la Réunion, Sainte-Clotilde, Réunion

Myriam Trémoulu École élémentaire publique Eugène Dayot, Saint-Pierre, France

Marc Troudet IREMI de Grenoble, Saint Martin d'Hères, France

Saranya Zearamane École Ormeteau, Aulnay-sous-Bois, France

Part I
Numbers and Calculations

Chapter 1
A Journey in Mayan Numeration

Alice Morales and Jean-Baptiste Meilhan

Abstract In this "Journey in Mayan Numeration", the research team "History in the teaching of mathematics" of the IREMI of Grenoble aims to provide beginners with the foundational concepts for understanding numbers through counting. The goal is to demonstrate that numbers are the result of a generative process, and that a numeration system is an organizational principle that allows the use of only a few symbols to represent numbers. To achieve this, this study explores the numeration system of the ancient Mayan, both spoken and written, within the context of pre-Columbian America.

Keywords Mayan civilization · Sexagesimal numeral system · Protraction · Classroom experimentation

Introduction

Deepening our understanding of a concept that is already somewhat familiar requires reigniting curiosity, posing new questions, and considering different perspectives. To truly grasp the workings and nuances of the decimal system (i.e., base 10) used for writing integers, it seemed both necessary and instructive to step outside of it in order to better "see" the objects and structures involved. In this chapter, we aim to explore and engage with numeration in a different base, using the positional vigesimal system (i.e., base 20) of the ancient Mayan civilization.

This approach feels natural, as it draws on historical observations that various numeral bases have been used throughout different cultures and time periods. Throughout history, not all societies have counted in base ten, and their motivations for adopting different systems have varied—ranging from trade to astronomy (including calendars and divination). In fact, some contemporary primitive societies,

A. Morales · J.-B. Meilhan (✉)
IREMI de Grenoble, Université Grenoble Alpes, Grenoble, France
e-mail: jean-baptiste.meilhan@univ-grenoble-alpes.fr

M. Moyon et al. (eds.), *Teaching Mathematics Through Historically-Based Activities*, History of Mathematics Education,
https://doi.org/10.1007/978-3-032-16990-7_1

isolated in remote areas (such as certain islands in Oceania or indigenous tribes in the Americas), still use body-based counting systems. For example, "five" is referred to as "one hand," "ten" as "two hands," and "twenty" as "one man." These spoken numerations preserve partial traces of base-5 or base-20 systems, remnants of a bodily conception of numbers (Keller, 2016).

The decimal system slowly gained ground in the West. In the tenth century, the monk (and future pope) Gerbert of Aurillac introduced it in his teaching at Reims, using Roman numerals to represent powers of ten and large numbers. However, it was not until the twelfth century that the numeral system we are familiar with—written with nine symbols and a zero, first invented in India centuries earlier—became more widely adopted in Western Europe. This system became known through Latin translations of Arabic treatises on "Indian calculus", as well as through works intended for merchants, which also provided algorithms for performing arithmetic operations. Today, we still retain legacies of these various numeral bases. Base 60, for instance, used in Mesopotamia long before our era, is still employed in measuring time. Traces of the base-20 system linger in the French language, such as in the way numbers between 80 and 100 are spoken (for example, 80 in French is literally "four-twenty"). These traces also persist in our cultural heritage, such as in the Quinze-Vingts ophthalmological center in Paris, which takes its name from the 300 beds it contained in the thirteenth century.

Choosing to immerse pupils in another culture also means making them aware that our humanity has been shaped by ancient civilizations, which had organized societies, cities, systems of education, and ways of living that were both similar to and distinct from our own. Moreover, the traces left by scholars—who have served as witnesses and gatekeepers of the past—offer us both a material and spiritual heritage, where numbers and calculations play a significant role, shaped by the needs of society. What may seem like mere instructions and duties to a schoolchild are thus deeply rooted in cultural contexts. In this way, modernity and progress take on a relative dimension, and forms of scientism are called into question.

Historical Context

To effectively engage in this immersive study of Mayan numeration, it is essential to first provide some historical, geographical, and societal context. Below, we offer key background information before turning our attention to Mayan writing and the economic role of cocoa, which will be explored through the proposed activities. For further reading, the works written by Coe and Kerr (1997), Coe (2015) and Restall (2020) can be consulted.

Geographical and Historical Contexts

The pre-Columbian Mayan civilization spanned a vast region of Central America, centered on the Yucatán Peninsula, which includes parts of present-day eastern Mexico, Guatemala, and Belize. The earliest evidence of human presence in the area—such as hunter-gatherer camps and Yucatán pottery—dates back to the second millennium BC. The true settlement of the lowlands began around 1000 BC. For this farming people, deeply tied to their land, the territorial boundaries remained largely unchanged over time and angles. This cultural and territorial continuity is a testament to the Mayan's remarkable ability to adapt and assimilate foreign influences. Today, approximately 3 to 5 million people still speak around thirty different Mayan dialects. The so-called "Classical" period of Mayan civilization, from 300 to 900 AD, marks its peak. This era is characterized by a flourishing of architecture, craftsmanship, population growth, and territorial expansion. However, it was also a fragile period. The land was not inexhaustible, prolonged droughts became more frequent, and agricultural techniques stagnated. The demands of an elite ruling class were heavy and diverse, contributing to social strain. These factors led to numerous conflicts and, by 900 AD, the collapse of the great Mayan cities. The Spanish conquest, beginning in 1540, ultimately brought an end to this civilization.

Society

The Mayan civilization did not have a centralized political power; instead, it was composed of autonomous city-states spread across a vast territory. However, some of these cities were true metropolises. For example, Tikal, in present-day Guatemala, had a population of around 50,000. Among the best-known Mayan cities today are Tikal, Chichén Itzá, and Palenque, all located in present-day Mexico and Guatemala.

The cities were typically organized around a central plaza. Dominating the space was a pyramid topped by a temple, which served both religious and funerary purposes. These pyramids often soared to several tens of meters in height, standing in stark contrast to the massive palaces, which were built horizontally and rarely exceeded ten meters in height. Other structures, such as steam baths for therapeutic and ritual use, ball courts, and platforms, were located nearby, all surrounding large open squares. The residential areas were arranged in a concentric pattern, extending outward toward the surrounding forests.

The political organization of the Mayan civilization was based on a hereditary monarchy. The ruler of each city held religious, military, and civil authority. The clergy formed a large and influential class, with priesthoods passing from father to son. Their responsibilities were diverse: they oversaw writing, chronology, the sacred almanac, medicine, the organization of ceremonies, and the education of future priests. Scribes and astronomers were highly specialized individuals whose roles extended beyond simply recording or transcribing the deeds of the kings. As diviners and priests, their tasks included interpreting and predicting events to safeguard the lives and futures of individuals, the community, and the dynasty. Writing was considered sacred, and astronomical calculations (cycles of Venus, the Moon, and eclipses) were primarily used for divination. These scholars, guardians of the sacred texts, are often depicted with the tools of their profession, such as engraving instruments or conchs used for calligraphy.

At the bottom of the social hierarchy were the common people, who were responsible for providing for the needs of the non-productive classes. Slaves, often drawn from common law offenders and prisoners of war, formed a distinct social class. It was frequently from among the slaves that sacrificial victims were chosen.

Writing

Archaeologists quickly recognized the phonetic nature of Mayan writing, and the numbers were deciphered relatively swiftly. However, understanding the texts proved more challenging, and it took nearly fifty years to fully decode the glyphic script. The Mojarra Stela, dating back to the second century AD, played a crucial role in this process, allowing scholars at the end of the twentieth century to crack the previously indecipherable writing. This stela features extensive glyphic texts that narrate the life of the monarch, his battles, and his familial relationships, making it a true "Rosetta Stone" for Mayan writing.

Mayan writing was employed on various materials, including stone and wood (such as stelae, lintels, and stairways), with the texts primarily glorifying royal power. Alongside the formal, intricate glyphic writing, a more cursive style also existed, where the designs were simplified. This cursive script appears on codices made from bark paper. These codices were essentially books made of bound pages and primarily dealt with astral sciences, such as astronomy, calendars, and divination. When the Franciscan missionaries overcame Mayan resistance in 1697, they not only massacred the population but, in their evangelical zeal, also destroyed idols and burned books. As a result, very few Mayan codices survived. Among those that did are the Dresden Codex (dating between 1240 and 1500, a copy of an earlier astronomical treatise), the Paris Codex (housed in the National Library of France, with similar dating), and the later Madrid Codex, created after the arrival of the Spaniards.

Cocoa

Cocoa was the prestige currency of pre-Columbian America (Rivero, 1990). According to historical linguistic sources, the Olmec civilization was the first to use cocoa as currency. Although cocoa itself served as a form of primitive currency and was one of the main mediums of exchange, the Aztecs and the Mayans primarily used pieces of cotton cloth, called *quachtli*, as a standard of value, representing a specific amount of labor. In the nineteenth century, Abbot B. de Bourbourg reported that cocoa beans continued to be used as small change among Guatemalan populations. In his *Grammar of the Quiché Language* (de Bourbourg, 1862), he explained that the number sixty *oxqal* represented three *ox* measures (*qal*). The *qal*, he noted, was the measurement of a cubit of cotton cloth, which had a common value equivalent to twenty cocoa beans. Thus, among the Mayans, cocoa became an integral part of social life through trade.

Scientific and Pedagogical Aspects

With this context in mind, we turn to how understanding Mayan numeration can deepen pupils' comprehension of numbers.

The aim of the proposed work is to provide pupils with the foundational concepts of number construction, rooted in counting and based on calculation. By involving them in this process, we aim to encourage pupils to recognize and engage with the generative process behind numbers. This principle allows only a few symbols to be used to unambiguously represent each number, enabling us to imagine numbers in their plurality and to extend them indefinitely (Caveing, 1990). A basic form of calculation, often based on addition and order (either orally or in symbolic form), emerges in the generation of numbers. It is through the practice of these first arithmetic operations that the structure of the numbering system begins to take root.

Mayan numeration is similar to our own decimal positional system, with the key difference that it is base twenty and incorporates the number five as an auxiliary base. Zero appears not only to indicate absence but also as a number in its own right.

The immersive approach, using a language and representations unfamiliar to our everyday experience, avoids any conversion exercises into our own numbering system. This approach encourages pupils—naturally curious at this age (9–11-year-olds)—to adopt a mindset of decoding and investigation. It becomes less about following familiar instructions and more about entering a new intellectual space to explore. At the end of the activities, and in light of usual classroom practice, pupils will be able to distinguish what is "the same" from what is "different." In this case, the term "system" will take on meaning through the recognition of a similar organizational structure in the scale of numbers, which dictates the same operational method for addition. What will be recognized as different are the notation, the writing system, and even the arrangements of the numbers. We hypothesize that the understanding of these practical actions, intended to become automatic, is deepened by the disorientation that accompanies the oralization of numbers and the manipulation of signs.

We first begin by outlining the main aspects of Mayan numeration. One can also refer to the texts written by Bidwell (1967), Lounsbury (1978), Closs (1968) and Anderson (1971) on this subject.

Mayan Numeration

A Matter of Time

The measurement of time is central to the construction of number in Mayan civilization. The codices, stelae, and monuments that have survived to this day testify to the important role played by "astral" sciences. These include the development of divinatory calendars, the expression of vast time spans, the tracking of celestial positions, and even the calculation of distant future dates, often tied to astrological beliefs. This emphasis on time led to the creation of a sophisticated system of time units. Without delving into the complexity of duration and date calculations, it is important to note that the Maya used two primary calendars: the *Tzolkin*, a sacred calendar of 260 days (comprising 13 months of 20 days each), and the *Haab*, a solar calendar of 365 days (comprising 18 months of 20 days plus 5 "unlucky" days). The *Haab* was roughly equivalent to a solar year. Additionally, the administrative year was counted as 360 days (18 months of 20 days). The combination of the *Tzolkin* and *Haab* calendars created a complex system where the same day in both calendars could only align once every 18,980 days, or 52 years. This period was known as a "century" or "old age," after which the cycle would begin anew.

Spoken Numeration

Intimately linked to the construction and structure of the Mayan language (more precisely, the Mayan languages), spoken numeration is based on a vigesimal system. To numbers from 1 to 19 (Table 1.1), are associated "unit classifiers" (Table 1.2)

Table 1.1 The numbers from 1 to 19

Hun	1	Uaxac	8	Ho-lahun	15 (five ten)
Ca	2	Bolon	9	Uac-lahun	16 (six ten)
Ox	3	Lahun	10	Uuc-lahun	17 (seven ten)
Can	4	Buluc	11	Uaxac-lahun	18 (eight ten)
Ho	5	Lahca	12	Bolon-lahun	19 (nine ten)
Uac	6	Ox-lahun	13 (three ten)		
Uuc	7	Can-lahun	14 (four ten)	Hun-kal	20 (one twenty)

Table 1.2 Some classifiers

(Hun)	-kal	-bak	-pic	calab	-kinchil
1	20	400	8000	160,000	3,200,000

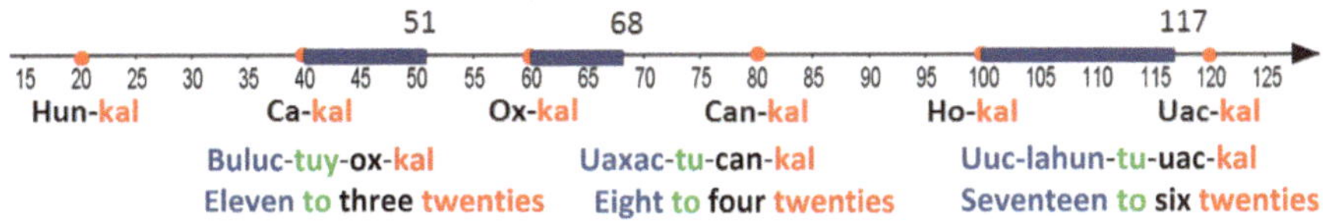

Fig. 1.1 Illustration of the oral construction of Mayan numbers

that designate bundles of units, bundles of twenty, bundles of twenty squared (400), bundles of twenty cubed (8000), and so on. In other words, these classifiers represent groups of 1; 20; 400; 8000; 160,000, etc. The parallel with our own system is clear: we use terms like hundred, thousand, and million to refer to bundles of 10^2, 10^3, or 10^6. These classifiers facilitate multiplication and the expression of multiples and powers of 20. For example, *Ox-kal* refers to three twenties (or 60), while *Ca-bak* refers to two four hundreds (or 800).

To express numbers that are not exact multiples of 20, the Maya used a particular system, and several hypotheses exist regarding their translation (see *Addendum* below). The dominant hypothesis in the literature since the 1980s is based on an "ordinal operation" known as "protraction" (Coe, 1982). This operation involves anticipating the next multiple of the base (a view of anteriority) and counting the complement from the previous multiple, effectively downgrading by one node (Cauty, 2002).

For our purposes, this would mean locating the number within a scale of values constructed from the base 20, its multiples, powers of 20, and their multiples. A graduated number line, like the one shown in Fig. 1.1, can help facilitate understanding.

For example, the number 68 lies between 60 and 80, so it is placed in the "4th twenties" (representing 60), and it is the 8th unit within this range. Therefore, it is expressed as *Uaxac-tu-ca-kal*, meaning "8 to eighty." Similarly, 51 is expressed as *Buluc-tuy-ox-kal*, meaning "11 to three twenties" (with *tu* becoming *tuy* when followed by a vowel).

Addendum: Recent work by Hoppan and Schwer (2021), published after the first edition of this book, questions the protraction principle proposed by Hagège for translating Mayan number-words. Hoppan and Schwer suggest that the system may instead be of a calendrical nature. According to their hypothesis, the number 51, *Buluc-tu-vox-kal*, would be translated as "eleven in the third twenties," and the number 68 would be rendered as "eight in the fourth twenties." However, the activity proposed in the context of our experiment (Section "Implementation in the Classroom") being based on the protraction principle, we were naturally led to retain this approach to spoken Mayan numeration in the current edition.

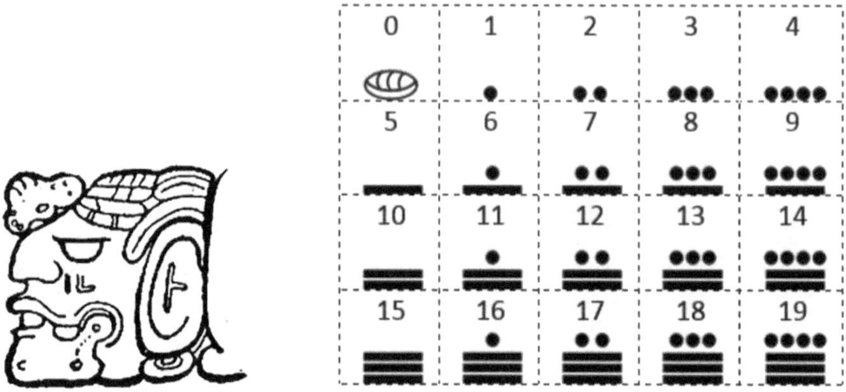

Fig. 1.2 (left) Example of glyph (here, number 13); (right) Numbers from 0 to 19 in "dot-bar"

Written Numerations

Written Mayan numerations use two forms of representation of numbers: the numeric glyphs and a symbolic numerical representation (the "dot-bar" symbols), see Fig. 1.2. Both representations can eventually be combined.

The "dot-bar" system uses only three signs: the dot, the bar, and an oval-shaped symbol representing zero (either a shell or an oval knife). To represent numbers from 1 to 20, the system is additive rather than positional, although there are regular patterns in the arrangement of dots and bars. This additive system, grouped in multiples of 20, is a base-20 numeral system written in colums. Depending on its vertical position (starting from the bottom), a dot represents a unit, a "twenty", a "twenty of twenties", and so on. Similarly, a bar represents five units, five twenties, and so forth.

For example, the number  written with our digits

10
9
0

is read, starting from the top: 10 twenties of twenties, 9 twenties and 0 unit. This "column" notation is always used with pupils; but this number can also be written in line as 10.9.0, which we will do in this text.

Pedagogical Perspective

Numeration forms the foundation for all subsequent elementary calculations—whether mental, reflective, or written—and serves as the first introduction to the mathematical world of numbers for young pupils. However, our teaching experience in Year 7 (11-year-olds) reveals that understanding decimal place value and performing basic arithmetic operations, skills that begin as early as Reception and continue throughout elementary education, presents significant challenges. Problems often arise with symbolization and the concept of position. For example, pupils frequently confuse "the digit of" with "the number of" (e.g., 3 tens and 40 units written as 340, or 2 thousands and 4 units written as 20004, or when 50 units are incorrectly considered greater than 7 tens). Grouping by class in large numbers is also often misunderstood (e.g., reading 2,343,254 as 234 325 4). A poor understanding of decimal notation also leads to difficulties with non-integer decimal numbers (often seen as just a juxtaposition of two integers), as well as in unit conversions (e.g., when converting 2.37 cm, pupils may place the 7 in the cm column of the conversion table). The institution also observes:

> For a pupil, the first steps in learning numeration are not as simple as they may seem. [...] The first learning experiences require the ability to designate numbers using a specific language (spoken) and within a universal system (written). [...] Additionally, to truly understand numbers, pupils must also internalize various language conventions that are influenced by cultural differences. [...] Finally, and most importantly, pupils must grasp that whole numbers can be written using just ten symbols (the ten digits: 0, 1, 2, ..., 9), and that the value of a digit depends on its position within a number. They must also understand the unique role of zero in this system. (CNESCO, 2015, p. 16).

Our proposed activities stem from the observation of the difficulties pupils face in understanding the principles of the positional decimal numeration system taught in our classrooms, with the aim of clarifying its meaning. The activities take time to step outside of established mechanisms or habits, encouraging pupils to reflect on and compare them with other systems. This is done through an immersive experience in Mayan culture, where pupils take on the roles of researchers—such as archaeologists, explorers, or translators—engaging with the construction of the Mayan positional numeration system.

The activities on Mayan numeration are structured around three key areas, covering the stages of the system's development: spoken numeration, symbolic and written numeration, base systems, positional writing, and basic calculations (addition and its iterations). We have divided the activities into three one-hour sessions, where pupils work in small, autonomous groups, followed by a collective review. The worksheets provided to the pupils are included in the appendix. Below is a brief outline of the structure and the goals of the activities:

1. Understand the generation of numbers in relation to one another, and how language reflects this relationship. Highlight the organizational principles behind number formation and the unique role of certain numbers, by deciphering unfamiliar numerical expressions in Yucatec, interpreting regularities and making conjectures. A collective discussion will follow, focusing on the oral expression of numbers in our decimal system and establishing comparisons.
2. Write numbers using the "dot-bar" system in base 20. Establish the positional nature of this system and examine its relationship to the practice of addition. Draw a parallel with our method of "carrying over." It is important to avoid any conversion to our decimal system, but to work directly within the base-20 system and understand how it functions.

3. Write and calculate with large numbers and highlight the powers of 20. Decrypt an excerpt from the *Dresden Codex*, specifically related to the cycles of the planet Venus: the multiples of 8.2.0 (found on page 24 of the codex).

Implementation in the Classroom

These activities were tested in five Year 7 classes (11-year-olds) across three secondary schools in the Grenoble district (France). To fully understand this section, it is recommended to refer to the pupil worksheets included in the appendix, as some instructions are explained orally with the support of a projector. For younger pupils, it is possible to initially focus only on the discovery of the "dot-bar" system and basic addition.

Session 1: Historical Introduction and Spoken Numeration

This session spans several disciplines: geography, history, French, and mathematics. It is divided into two parts and involves extensive collective interaction.

As an introduction, the activity begins by placing the Mayan civilization in space and time. Some key elements of Mayan civilization are presented to the pupils through a slideshow. The aim is to give substance to the concept of "civilization" and to humanize it. Over long periods, civilizations rise and fall, leaving behind traces that make the past a subject of study—traces that also include everyday objects of human life. History is not just told; pupils actively engage in its discovery. The Mayan territory is first located on a world map, followed by a more detailed map of Central America. The geographical features of the region are highlighted, including the mountainous terrain on the Pacific coast and the tropical forests in the center and on the Atlantic coast. The teacher also names the modern countries that occupy this territory—Guatemala, Belize, eastern Mexico (including Yucatán), parts of El Salvador, and Honduras—and outlines the land occupation, noting the locations of major Mayan cities. This presentation is followed by an illustration of Christopher Columbus's ships. The teacher then prompts the pupils with a question about the date of the discovery of the Americas by Europeans.

Next, a short excerpt from the documentary *Le code maya enfin déchiffré* [The Mayan Code Finally Deciphered] (Lebrun, 2016) is shown, which presents some of the surviving Mayan artifacts (monuments, pottery) and introduces the pupils to the complex Mayan glyphs. This first encounter with Mayan writing offers an opportunity to discuss the long process of deciphering these glyphs. The teacher highlights that in discovering the American continent, the Spaniards also encountered and introduced many plants still present in modern diets, such as corn, tomatoes, and potatoes. One of these plants, cocoa, played a central role in

Mayan civilization. It was prepared as a beverage known as "the drink of the Gods," reserved for elites such as kings, priests, and warriors. Cocoa beans were also used as currency and a means of payment.

After this 25-minute historical introduction, the remainder of the session is dedicated to exploring Mayan numbers and "spoken numeration". Pupils are given worksheets, which are also presented as slides for collective work. Numbers from 1 to 12 written in Yucatec are introduced. The teacher explains that the "u" is pronounced [u] and invites several pupils to read the numbers aloud. They are then asked to translate these expressions into French in the central column of their worksheet and identify the additive structure underlying the Yucatec number words. The pupils quickly grasp the structure of numbers from 12 to 19. For example, *Can-lahun* is composed of Can (4) and *Lahun* (10). The number 20 introduces a new word: *Hun-kal* (one-twenty). The pupils then naturally ask, "And what about 21?" Without providing an immediate answer, the teacher places the following multiples of 20 on a graduated half-line: Ca-kal (two-twenty), *Ox-kal* (three-twenty), *Can-kal* (four-twenty), and so on, up to *Bolon-lahun-kal* (nine-ten-twenty). For 20-twenty, a new word appears: *Hun-bak*.

Writing the numbers on a graduated half-line helps the pupils visualize the important "nodes" of the numeration system. When the teacher reaches the position of 30 on the line, the writing is shown as *Lahu-ca-kal*. Some pupils are surprised by this form and suggest an interpretation of "40 minus 10." The teacher then clarifies that 21, for example, is written as *Hun-to-ca-kal*, which contradicts the earlier hypothesis (since *Hun-to-ca-kal* would be interpreted as "40 minus 1"). In fact, 21 is written as "one in the second twenty," meaning *Hun-to-ca-kal* translates as "one in the second twenty."

The pupils then practice with other examples. While this step may prove difficult for some, most pupils come to understand the central role of the number twenty in the Mayan numeration system. By the end of the session, they have recognized the regularity in the way numbers from *Hun* to *Hun-bak* are expressed. The teacher reformulates these regularities and, through interaction with the pupils, compares them to the enunciation of numbers from one to a hundred in the French decimal system.

Session 2: "Dot-bar" Numeration

This second stage aims at discovering the "dot-bar" writing of Mayan numbers and then the vigesimal numeration. The goal is to deepen pupil's understanding through the practice of operations. The calculations are carried out for now up to 19 twenties 19 units.

At the start of this session, the teacher draws a graduated half-line on the board, ranging from 0 to 600. The teacher then asks the pupils to identify how numbers like 1, 3, 20, 60, and 400 are written in Yucatec. The corresponding written forms of

Fig. 1.3 Illustration of grouping

these numbers remain displayed on a portion of the board for reference. Following this introduction, the teacher then proposes the following activity.

> During the cocoa pod harvest, Tiso collected beans and made two piles, then he noted the number of beans he had gathered.
>
> 
>
>
> a. What could the symbols — and • possibly correspond to?
> b. Tiso has grouped all his beans into one pile. In your opinion, how will Tiso write the total number of beans using the two symbols?

Dot is easily associated with a unit, and the bar with five units. The symbolic representation of numbers from 1 to 19 naturally follows from this system. This provides an opportunity for the teacher to introduce the glyphs corresponding to these numbers, which often surprises the pupils.

To practice these numbers, the pupils then perform additions that help introduce the concept of writing numbers from bottom to top, as well as the use of zero. In the first addition exercise, pupils group the dots and bars without converting them into the decimal system (see the color coding in Fig. 1.3).

In the second addition (12 + 8), pupils begin by writing four bars to represent twenty. The concept of positional writing in "layers" (or "floors") is then introduced, along with the symbol for zero (the knife or oval), linking it to the *Kal* in spoken numeration (from session 1). The representation of 20 is then reinforced, and pupils practice completing numbers up to 20 to solidify their understanding of the two-level positional system, as illustrated in Fig. 1.4.

The example in Fig. 1.4, where the error is identified through the addition process, requires a clear explanation of the rules for composing "dot-bar" numbers. Once these rules are understood, the pupils successfully complete the additions up to 20, as demonstrated in Fig. 1.5.

Then, additions of increasing difficulty are proposed, initially focusing on the class of twenties. These activities are carried out independently. However, this

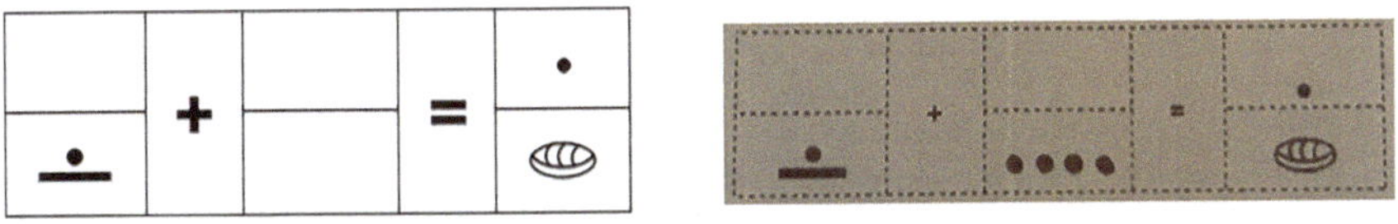

Fig. 1.4 Example of statement and an incorrect answer

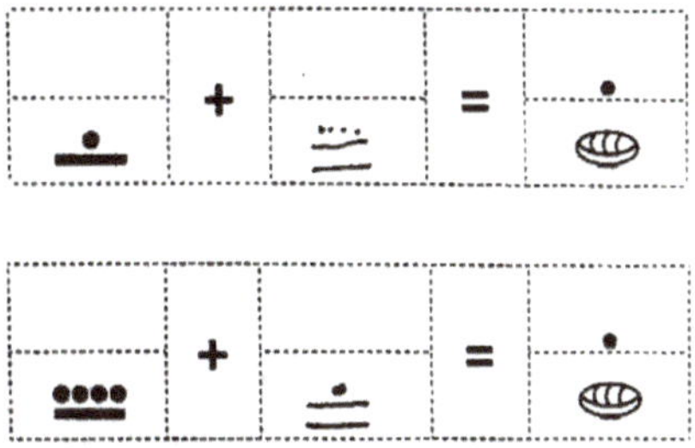

Fig. 1.5 Complement to twenty. The instruction is "Complete the following sums"

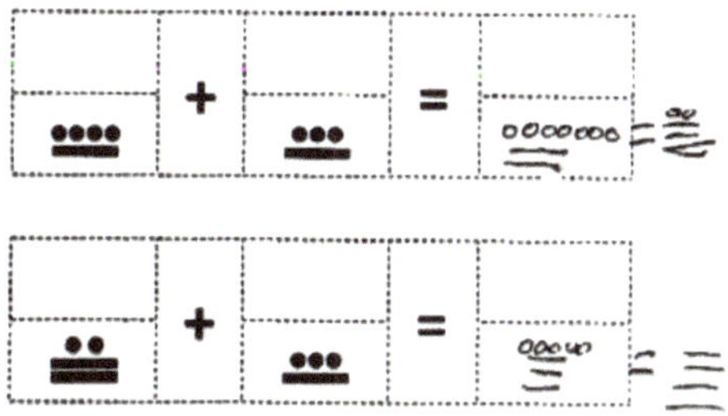

Fig. 1.6 Half-success, the writing of 20 is not understood

Fig. 1.7 Examples of errors arising from the use of the decimal system

writing system is not naturally acquired by all pupils. In Fig. 1.6, for example, the pupil understood that 5 dots should be replaced by a bar, but has not yet grasped that 4 bars should be replaced by a dot in the next higher class.

The conversion between "dots-bars" writing and decimal notation, and vice versa, is another common source of errors. Two examples are shown in Fig. 1.7. In the example on the left, the pupil begins by converting the "dots-bars" notation into decimal form to perform the addition but then mistakenly reconverts it back into "dots-bars" without leaving the decimal system. The example on the right contains two errors. First, the addition itself is incorrect, as it combines the two writing systems. Second, the conversion of the result back into "dots-bars" is also erroneous. In both cases, the empty boxes in the table format may have led the pupils to revert to decimal writing.

During the collective correction of some of these additions, the Yucatec reading of certain numbers is revisited to further highlight the role of the term *Kal*. At the end of the session, a pupil is invited to write a new addition on the board. The entire class successfully completes the operation, except for a few pupils.

Session 3: Study of the *Dresden Codex*

This third session is divided into two parts: the first part focuses on additive calculations involving multiples of twenties, leading to a three-level (or "floor") representation and transcription using our numerals. The familiarity gained from the "dots-bars" additions allows the second part to be conducted independently, with pupils working on the *Dresden Codex*.

A new addition is introduced (Fig. 1.8), which brings in the higher level—the *twenties of twenties*—linking it with the *Bak* from spoken numeration. The pupils then complete several similar additions.

At this stage, the notation with our numbers is introduced.

The number 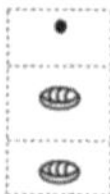 is denoted by

1
0
0

. Care is taken to verbalize the results of these operations: the number

1
13
2

is read "1 twenties of twenties, 13 twenties, 2 units".

The teacher then introduces the *Dresden Codex*, available on https://digital.slub-dresden.de/werkansicht/dlf/2967/1, a 39-page manuscript made of vegetable paper, painted on both sides, and written in the Yucatec Mayan language. Discovered in the nineteenth century in the region of Chichén Itzá, this codex provides context for the film excerpt shown during the first session. It contains a wealth of astronomical calculations, and the focus of this activity is on a section from page 24, which deals specifically with the planet Venus. This section includes only the first eight multiples of 8.2.0, derived through repeated additions (bottom right of the page).

The pupils are tasked with deciphering the numbers written at the bottom of the page and hypothesizing a relationship between them. They quickly identify a doubling pattern between the first two numbers, 8.2.0 and 16.4.0. But what about

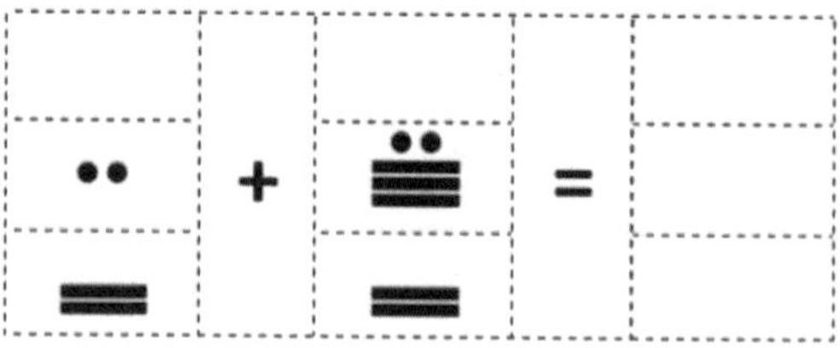

Fig. 1.8 Example of an addition

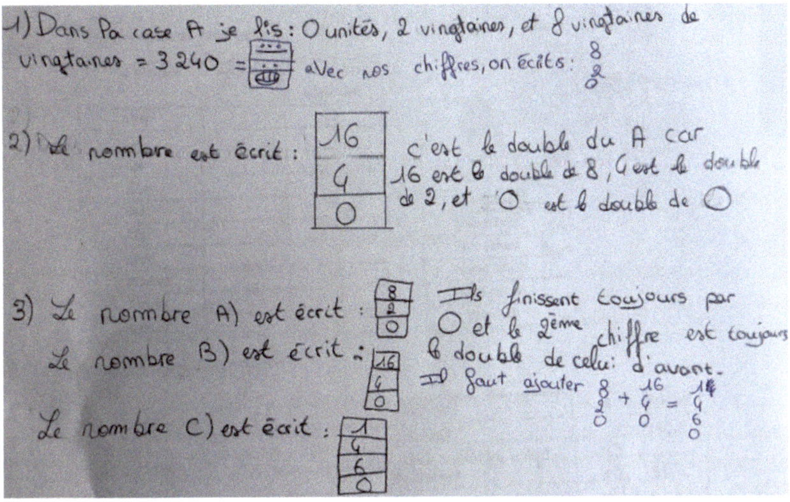

1) In box A I read: 0 units, 2 twenties, and 8 twenties of twenties = 3240 =  with our numbers, we write 8 | 2 | 0

2) The number written is 16 | 4 | 0 It's the double of A because 16 is the double of 8, 4 is the double of 2, and 0 is the double of 0.

3) The number A is written 8 | 2 | 0 They always end with 0 and the second digit is always double of the one before.

The number B is written 16 | 4 | 0 We need to add 8 | 2 | 0 + 16 | 4 | 0 = 1 | 4 | 6 | 0

The number C is written 1 | 4 | 6 | 0

Fig. 1.9 Column writing with our numbers is mastered (with English translation)

the others? The more "intuitive" pupils suggest the numbers may follow a pattern of successive multiples. They then verify this hypothesis through calculation. An error made by the scribe in one of the multiples captures their attention, as does the distinctive notation for zero, represented by the shell symbol. Writing with our numbers appears to have significantly reduced the errors, as long as the pupil understood that the largest number they could write was 19 (see Fig. 1.9).

In classes where work on Mesopotamian numeration had been conducted beforehand, some pupils drew parallels between the two systems. For example, we heard remarks like, "The shell is practical; it helps prevent you from getting the number wrong," or "They too had just two symbols."

Ways to Go Further

Some Observations

It is difficult to evaluate this experience comprehensively over such a short time, but we would like to highlight a few key points:

- This type of approach, rich in historical narratives that can be developed according to the teacher's interests and cultural background, fosters pupil involvement and motivation.
- The pupils quickly grasped the meaning of the dot (representing one unit) and the bar (representing five units). However, immersing pupils in a numerical system that diverges from their habitual decimal system was more challenging. The number is so intrinsically tied to its decimal representation that the idea of a different system sometimes seemed nearly impossible for some pupils to grasp. This led, for some, to hybrid representations where both decimal and "dots-bars" systems were used together (sometimes humorously, as seen in Fig. 1.7). This highlights the intellectual challenge of separating a number from its conventional written representation.
- Using an unfamiliar language proved beneficial. The deciphering of numbers in Yucatec Maya was relatively easy for the pupils. This activity helped them realize that numbers are structured in a sequence, with language reflecting this structure. The principle of "protraction" (explained earlier) posed some challenges, but it also provided a fresh perspective on our own numbering system. In essence, naming a number in the Mayan system corresponds to locating it within a geometric framework: placing it unambiguously on a "scale" constructed from the base, identifying the interval in which it belongs, and pinpointing its position within that interval.
- Pupils also became aware that numeration is constructed based on powers of a base. For example, the relationship between 100 and 10 in our decimal system mirrors that between *Hun-Bak* (400) and *Hun-Kal* (20) in the Mayan system.

Details on Units of Time

In the Mayan metrological system, units of time follow a largely vigesimal (base-20) structure: 1 day (*kin*), 20 days (*uinal*), 1 year (*tun*), 20 years, 400 years, etc. Except in one case, each unit of time is 20 times the previous one. The only exception is the administrative year (*tun*), which consists of 360 days or 18 months of 20 days. In Mayan representations, the distinction between durations and dates is made. The numbers found on page 24 of the *Dresden Codex*, for example, represent durations. In this case, 8.2.0 corresponds to 8 years, 2 months, or 2920 days—equivalent to five synodic revolutions of Venus, each lasting 584 days. We deliberately chose to focus on the first eight multiples of 8.2.0, where the calculations follow the vigesimal system, as the method of calculation involves repeated additions. At the ninth multiple, however, an anomaly occurs: adding 8.2.0 to 3.4.16.0 results in 3.13.0.0 instead of 3.12.18.0, indicating an error in the codex. An extension of the activity could involve studying the remainder of the codex up to the thirteenth multiple, which delves deeper into the complexities of Mayan calculations, especially as they relate to time and the religious significance of the Mayan calendar (Cauty, 1998).

Possible Extensions

We believe that the real value of the activities proposed here lies in their immersive approach into a culture and system of thought. Rather than just presenting the novelty and appeal of new forms of writing or speech, this approach goes deeper. By practising these systems, pupils gain a better understanding of the mechanisms behind numeration. This immersive experience can help them see numeration not as an isolated system but as something deeply embedded in the culture and history from which it originates. This is where the true power of historical approaches can influence pupils' understanding of our own numeration system.

Notably, our approach avoids exercises focused solely on conversions (a common feature in some literature), where historical numeration systems are merely used as a pretext to practice modern mathematical operations. While a broad survey of historical numeration systems is not necessarily a valuable exercise on its own, it would be fruitful to extend the activities proposed here by conducting a comparative study. By comparing the Mayan system with others, pupils can better appreciate the advantages and limitations of the Mayan system.

For instance, the Egyptian numeration system is well-known and commonly featured in textbooks. It could be interesting to explore its particularities, such as its non-positional nature and its limitations when working with large numbers.

The Mesopotamian system offers much richer historical sources, especially in terms of mathematical writings. This system's sexagesimal (base-60) structure offers fascinating features, including floating values in positional notation, which could provide an additional layer of insight for pupils. We recommend the brochures for further exploration of these topics (Morales et al., 2015, 2016).

Another extension could involve exploring the Chinese numeration system, particularly its associated computational techniques such as abacus calculations or counting rods. Our experience shows that once pupils have engaged with a different numeration system, they find it easier to approach and compare others. A comparative approach to numeration systems not only deepens their understanding but also naturally promotes an appreciation for the diversity and ingenuity of historical mathematical practices.

Acknowledgments This chapter was written by Alice Morales, Jerome Capitan, Ludovic Jollet, Anne Jorioz, Anne-Marie Marmier and Jean-Baptiste Meilhan. All these people worked together within the research team "History in the teaching of mathematics" of the IREMI of Grenoble (France).

Appendix 1: Concepts and Skills Addressed in This Chapter

Covered Concepts

- Use and represent large whole numbers
  - Compose, decompose large whole numbers, using groupings by thousands (units of numeration: simple units, tens, hundreds, thousands, millions, billions) and their relationships.
  - Understand and apply the rules of numeration to large numbers up to 12 digits.
- Calculate with whole numbers
  - Set calculation: implement a set calculation algorithm.
  - Operational calculation techniques.

Developed Skills

- **Search and research skills**: Extract, organize information, manipulate, experiment from a new support.
- **Reasoning**: Perform calculations requiring the construction of a process that combines reasoning steps (group to better count, group-exchange to calculate).
- **Computational skills**: Use a new operational technique in connection with classic techniques.
- **Mathematical communication skills**: Explain one's approach and argue in exchanges.
- **Mathematical representation ability**: Produce different representations of a number.
- **Modeling**: Recognise an additive and multiplicative situation.

Appendix 2: Pupil Worksheet Used in Year 7

Available at
https://www.univ-irem.fr/chapitre-1-voyage-en-numeration-Maya

The Mayan Numeral System

1. **Exercise 1**

   (a) Numbers in Yucatec 1
   Using the table on the left, complete the analysis of the table on the right.

Yucatec	Decimal notation
Hun	1
Ca	2
Ox	3
Can	4
Ho	5
Uac	6
Uuc	7
Uaxac	8
Bolon	9
Lahun	10

Yucatec	Analysis	Decimal notation
Buluc		11
Lahca		12
Ox-lahun		13
Can-lahun		14
Ho-lahun		15
Uac-lahun		16
Uuc-lahun		17
Uaxac-lahun		18
Bolon-lahun		19
Hun-kal		20

   (b) Numbers in Yucatec 2

      (i) Complete the table

Yucatec	Analysis	Decimal notation
Hun-kal	One-twenty	
Lahun-ca-kal		
Ca-kal	Two-twenty	
Lahun-ox-kal		
Ox-kal	Three-twenty	
Lahun-can-kal		
Can-kal		
Lahun-tu-yo-kal		
Ho-kal		

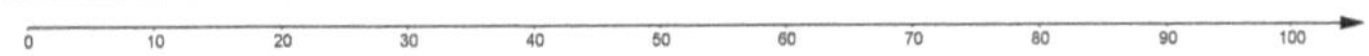

(ii) Complete the table

Yucatec	Analysis	Decimal notation
Ho-kal		
Uac-kal		
Lahun-kal		
Ho-lahun-kal		
Bolon-lahun-kal		
Hun-bak		
Ca-bak		

(c) But how do you say 21, 22, etc.?

(i) Using the example from the first line, complete the following table:

Yucatec	Analysis	Decimal notation
Hun-tu-yox-kal	One-to-three-twenty	41
Buluc-tu-yox-kal		
Uaxac-tu-can-kal		
Uuc-lahun-tu-uac-kal		

(ii) Place the numbers from the table on the following graduated half-line:

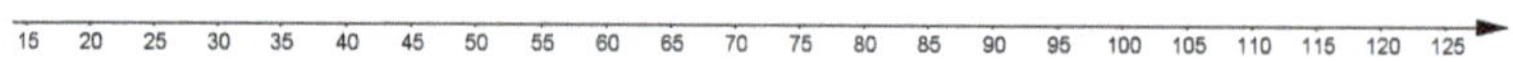

2. **Exercise 2**

During the cocoa pod harvest, Tiso collected beans and made two piles, then he noted the number of beans he had gathered.

(a) What could the symbols ▬ and • possibly correspond to?

(b) Tiso has grouped all his beans into one pile. In your opinion, how will Tiso write the total number of beans using the two symbols?

Summary:

....

....

3. **Exercise 3**
   Perform the following additions in "dot-bars":

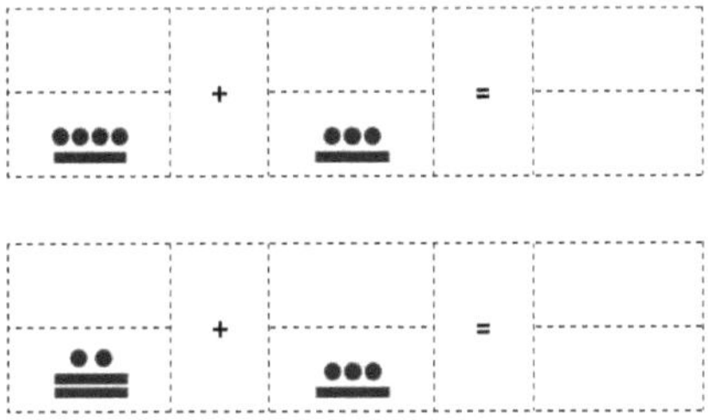

   Summary:
   ....
   ....
   ....

4. **Exercise 4**
   Perform the following additions in "dot-bars":

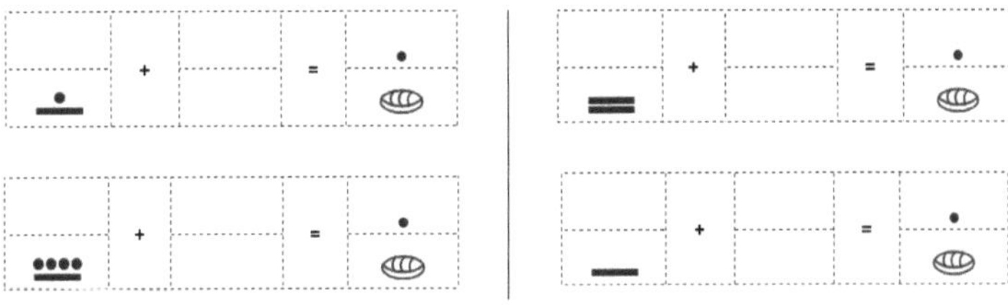

5. **Exercise 5**
   Perform the following additions in "dot-bars":

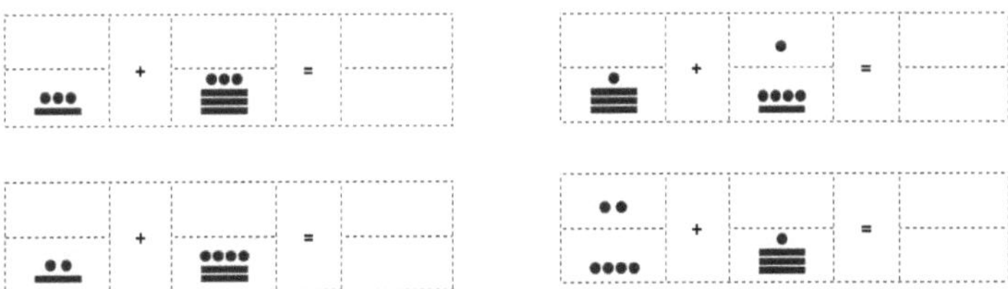

6. **Exercise 6**
   Perform the following additions in "dot-bars":

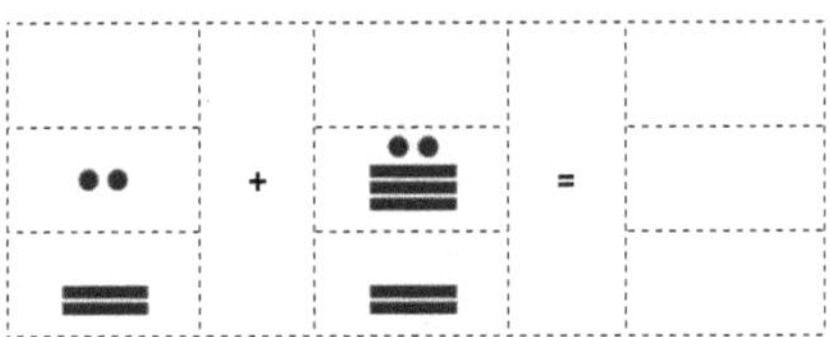

Summary:

....

....

....

7. **Exercise 7**

   Here is a part of page 24 of the Dresden Codex. This codex contains astronomical calculations, some of which concern the planet Venus.

   (a) What is the number written in a)?
   (b) What is the link between the number written in b) and the number written in a)?
   (c) What link exists between the numbers written in a), b) and c)?
   (d) Continue this work up to h).

Extract from the Dresden Codex, kept at the Library of the University of Saxony, Dresden, Germany.
Denomination of columns, IREM Grenoble.

Copy of the Dresden Codex, William Gates, The Dresden Codex, The Maya Society at the Johns Hopkins University, Baltimore, 1932.
Denomination of columns, IREM Grenoble.

References

Anderson, F. (1971). Arithmetic in mayan numerals. *American Antiquity*, *36*(1), 54–63.

Bidwell, J. (1967). Mayan arithmetic. *Mathematics Teacher Learning and Teaching PK-12*, *74*, 762–768.

Cauty, A. (1998). Lire et faire parler un texte: par qui et comment les pages 24 à 29 du Codex de Dresde peuvent-elles être traduites? *Amerindia*, *23*, 1–26.

Cauty, A. (2002). Le type protractif des numérations de l'aire maya. *Faits de langues*, *20*(1), 85–93.

Caveing, M. (1990). Quelques précautions dans l'emploi de l'idée du nombre. *L'Homme*, *30*(116), 151–157.
Closs, M. (1968). The mathematical notation of the ancient mayan. In M. Closs (Ed.), *Native American mathematics* (pp. 291–370). The University of Texas Press.
CNESCO. (2015). *Nombres et opérations : premiers apprentissages à l'école primaire*. CNESCO.
Coe, M. (1982). *La structure des langues*. Presses universitaires de France.
Coe, M. (2015). *The Mayan*. Thames and Hudson.
Coe, M., & Kerr, J. (1997). *The art of the Mayan scribe*. Harry N. Abrams.
de Bourbourg, A. B. (1862). *Gramatica de la Lengua Quiche*. A. Bertrand.
Hoppan, J., & Schwer, S. (2021). De la protraction en maya ? *Faits de langues*, *51*(2), 39–62.
Keller, O. (2016). L'invention du nombre. In L. Radford, F. Furinghetti, & T. Hausberger (Eds.), *History and pedagogy of mathematics: July 18–22 2016, Montpellier. Proceedings of the 2016 ICME Satellite Meeting* (pp. 703–712). IREM de Montpellier.
Lebrun, D. (2016). *Le code maya enfin déchiffré*. Arte.
Lounsbury, F. (1978). Mayan numeration, computation, and calendrical astronomy. In C. Gillispie (Ed.), *Dictionary of scientific biography* (Vol. 15, pp. 759–818). Scribner's.
Morales, A., Berthier, N., Capitan, J., Jollet, L., Jorioz, A., Marmier, A.-M., Meilhan, J.-B., & Racine, C. (2016). *Les Mathématiques en Mésopotamie & Variations sur les aires – Niveaux Collège et Lycée*. IREM of Grenoble.
Morales, A., Capitan, J., Jollet, L., Jorioz, A., Meilhan, J.-B., & Racine, C. (2015). *Les Mathématiques en Mésopotamie – Niveaux 6ème et 5ème*. IREM of Grenoble.
Restall, M., & Solari, A. (2020). *The Mayan: a very short introduction*. Oxford Academic.
Rivero, P. (1990). When cocoa was used as currency. *The UNESCO Courier*, *XLIII*(1), 16–19.

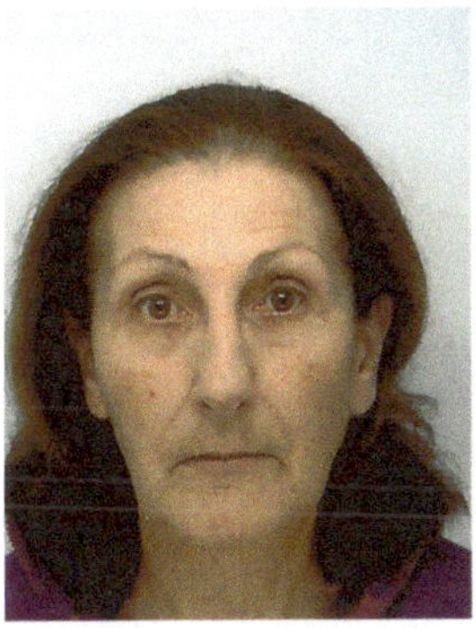

Alice Morales is a retired mathematics teacher who worked at a middle high school in Saint-Martin-d'Hères, near Grenoble. She co-founded in 2009 the "History and Teaching of Mathematics" group at the IREM (Institute for Research and Education in Mathematics) in Grenoble, which she directed until 2019.

Jean-Baptiste Meilhan is an associate professor in mathematics at the Institut Fourier, University of Grenoble Alpes. His research focuses on low-dimensional topology, with a particular interest in knot theory. For over ten years, he has also been a member of the "History and Teaching of Mathematics" group at the IREM (Institute for Research and Education in Mathematics) in Grenoble. After working on Mesopotamian and Mayan mathematics, the group has spent the past few years studying the mathematics of ancient China. Their work is documented on a blog titled "Neuf Chapitres Trois Quarts" (https://neufchap.hypotheses.org/).

Chapter 2
From the Token Abacus to Written Calculation

Nathalie Daval and Dominique Tournès

It is however very true that the abuse of words is a great source of errors, for there comes a kind of calculation error, as if in calculating one did not mark the place of the token well.

Gottfried Wilhelm Leibniz (1765, p. 217)

Abstract Token abaci were one of the main calculation tools in the West until the end of the eighteenth century. They were gradually replaced, not without difficulty, by written calculation methods from India using Arabic numerals, before the latter were gradually automated using mechanical and then electronic calculators. The crucial role played by token abaci in the history of calculation led us to question the relevance of using these ancient material artefacts to develop or consolidate basic numerical skills. After a brief historical presentation on token abaci, the chapter reports on some educational experiments that were carried out in France in primary school classes (pupils aged 7–9) over three school years. The common feature of the pupils' activities was the use of historical token abaci to learn the decimal positional system and the associated calculation techniques. These experiments led to the design of models of token abaci adapted to various levels and to the construction of the teaching progressions detailed here.

Keywords Token abacus · Written calculation · Decimal positional numeration · History of calculation tools · Remediation to calculation difficulties

In order to appropriate the decimal numeration of position and the basic operational techniques, it seems indispensable that numbers and calculation are instrumented by different things (fingers, abaci, string with knots, pebbles, tokens, fictitious money,

---

N. Daval · D. Tournès (✉)
Parc Technologique Universitaire, LIM, IREMI de la Réunion, Université de la Réunion, Sainte-Clotilde, Réunion
e-mail: nathalie.daval@umontpellier.fr; dominique.tournes@univ-reunion.fr

M. Moyon et al. (eds.), *Teaching Mathematics Through Historically-Based Activities*, History of Mathematics Education,
https://doi.org/10.1007/978-3-032-16990-7_2

graduated ruler, calculator...), and that children manipulate tangible objects for a long time before moving on to symbolic representations.

This chapter is focused on one of the most frequently used calculation aids in the past: the token abacus. We can agree to use the generic term "abacus" to designate all the instruments in which small identical objects (tokens, pebbles, balls, shells...) are placed on columns, lines or parallel rods materialised on a flat support, in order to represent numbers and to calculate on them. The word "abacus" comes from the Greek *abax* and the Latin *abacus*, one of whose meanings was "counting table". In the large family of abaci, we distinguish two categories of instruments: on the one hand, the "token abaci", mainly used on the western side of the Old World, whose principle is to place tokens on a calculating table marked with lines or columns, and, on the other hand, the "bead abaci", especially used on the eastern side, which consist of sliding beads on parallel rods fixed in a frame. The major difference between these two families of artefacts is that, in the token abaci, one can place an unlimited number of tokens on each row or each column, while in the bead abaci, the number of beads that can slide on a given rod is limited by construction.

The abacus was the main calculation instrument of humanity for over 2000 years (Pullan, 1969; Sugden, 1981). In this chapter, we will no longer talk about bead abaci, but only about token abaci (Schärlig, 2001, 2006). This chapter is based on the belief that the use of these ancient artefacts, deeply rooted in our culture, presents a didactic interest in promoting pupils' learning in the numerical field. After recalling the historical context of the appearance and development of token abaci (Tournès, 2016), we will detail some ideas for their use in the classroom.

Historical Context

At the Sources of the Token Abacus

Archaeology has found fragments of about thirty abaci in the Greek world. These are most often engraved stones located at the edges of monuments, which probably served as calculation tables for money changers, bankers and merchants involved in the activities of the temples. The oldest known abacus, dated from the fifth or fourth century BC, was discovered on the island of Salamis, not far from Athens. It presents itself in the form of a white marble slab 149 cm long and 75 cm wide, bearing parallel lines. In addition to the lines, are engraved on the abacus signs representing the monetary units and the acrophonic numbering system used by the Greeks. It is a base 10 system, but with a sub-base 5, similar to the one that was adopted later by the Romans (Fig. 2.1).

We encounter numbering systems of the same type as the Greek acrophonic system in most civilizations and at all times. The base 10 is most likely derived from our ten fingers. Similarly, the sub-base 5 evokes a hand with five fingers, but it is probably also an inevitable consequence of the fact that the human eye is not

Fig. 2.1 The Greek acrophonic numeration (Schärlig, 2001, p. 44)

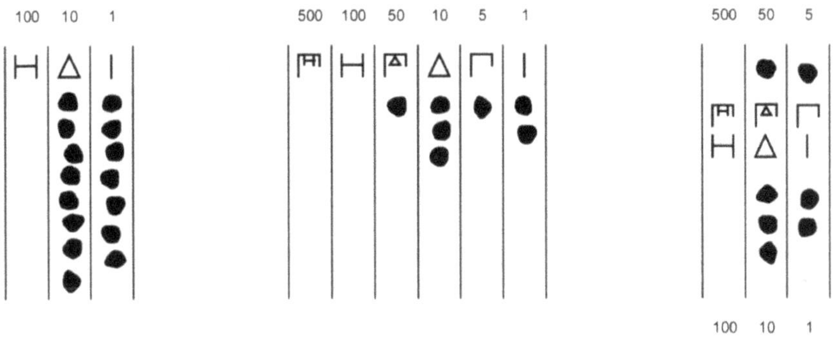

Fig. 2.2 Three ways to inscribe the number 87 on Greek abaci (Schärlig, 2006, p. 18)

capable of perceiving globally and instantly a number of objects beyond four. This phenomenon, called "subitizing", has been studied and explained quite convincingly by neuroscience (Dehaene, 1997). Newborns and many animal species are capable of discriminating a small number of objects, which shows that it is an ancestral ability that appeared very early in evolution. On the other hand, from five objects, the human being must mentally decompose the set of objects into two or several subsets of fewer than five elements. It is therefore probably because of the neurological phenomenon of subitizing that, in most ancient numerations, numbers from 1 to 4 were represented by a vertical line (or any other sign) repeated identically, then something was invented from 5 to avoid counting, either by the introduction of new abbreviating signs, or by the arrangement of elementary signs in easily recognizable geometric constellations. For example, in Mesopotamia, we observe the grouping of signs by three before moving to the next line.

To return to the Greek abaci, their analysis has shown that there were probably three ways to arrange the pebbles on the calculating surfaces, the most common (in the centre of Fig. 2.2) consisting of arranging the columns in direct correspondence with the written numeration. In the current vocabulary of abaci, the pebbles or other objects that represent 1, 10, 100, etc., are called "unaries" and those representing 5, 50, 500, etc., are called "quinaries". The use of quinaries ensures that there are never more than four objects in each column or half-column, which facilitates a direct perception of the numbers inscribed on the abacus (especially in the arrangement on the right, which is later found in most abaci with a central bar, the quinaries being located above and the unaries below).

Following the Greeks, the Etruscans and then the Romans used token abaci. The best-known document on this subject is the one known as the gem of the Etruscan calculator: preserved in the *Cabinet des Médailles* in Paris, of uncertain date, it is an engraved stone on which a man is depicted calculating with tokens on his counting table and recording the results of his calculations on a wooden tablet bearing Etruscan numbers. As for the Romans, we have several written sources of their abaci, as well as various archaeological testimonies. The most interesting is undoubtedly the bas-relief adorning a Roman sarcophagus from the first or second century, preserved in the Capitoline Museum in Rome, which shows us a young calculator standing, performing an arithmetic operation on an abacus in front of his master and mistress. The *calculators* or *numerarii* were professionals responsible for teaching the art of calculation and keeping accounts in the large houses of the patricians.

From the Token Abacus to Indian Calculation

In the West, the practice of calculating on the token abacus persisted at least until the end of the eighteenth century. Treatises on calculation with tokens flourished throughout this period. For example, in *L'Arithmetique de Jean Trenchant, avec l'art de compter aux Getons* (1617), we observe that the unaries are placed on parallel horizontal lines and the quinaries between the lines, which then became the most common arrangement. This is the one we have adopted for our educational experiments.

To understand this longevity of the token abacus, it must be said that Roman numeration, like the Greek acrophonic numeration that preceded it, was not easily adapted to calculation: Roman numerals could certainly be used to write numbers, but in no case to operate on them. Moreover, keep in mind that in Europe until the eighteenth century, very few people knew how to write outside of those moving in elite circles, which, in addition to the fact that paper was very expensive, did not facilitate the spread of the technique of Indian calculation with the zero and Arabic numerals, although transmitted quite early to Europe by the intermediary of Muslim Spain (*Andalus*). Indeed, we know today that the so-called Arabic numerals and zero arrived in the Iberian Peninsula in the tenth century (at least one manuscript witness has been found) and that, from the twelfth century, there was a whole literature related to Indian calculation (called *algorism*) (Evans, 1977).

Indian calculation, later called "pen calculation", is none other than the "written calculation" that we learn today in school. It is a particular instrumented calculation based on the paper-pencil artefact, which tends to do without the support of the abacus by noting on paper, in a symbolic way, the manipulations to be performed. We find its trace in the alignment of the numbers in fictitious columns mimicking those of the abacus, in the decomposition of operations into partial operations involving the different digits of numbers separately, and in the exchanges that are made: exchange of ten units for a ten, which is translated into a carryover, reverse exchange of a ten for ten units to make a subtraction or a division possible. Moreover, many old multiplication and division algorithms, in which intermediate partial results are successively written and crossed out, better preserve the memory of these concrete manipulations than ours do.

While European scholars—foremost among them Leonardo of Pisa, known as Fibonacci (Moyon, 2016)—enthusiastically adopted the new written calculation that contributed to the rise of mathematics in the West, merchants and bankers continued to use the abacus with tokens (Smith, 1921). The same was true for individuals to carry out simple daily life calculations, as evidenced by this excerpt from a letter from Madame de Sévigné to her daughter dated 10 June 1671: "We have found, with these tokens that are so good, that I would have had five hundred and thirty thousand pounds of property, counting all my little inheritances". The quarrel was long and fierce between the *abacists* (supporters of the abacus with tokens) and the *algorists* (supporters of written calculation), as subtly evoked by the famous engraving of 1503, taken from the book *Margarita Philosophica* by Gregor Reisch. The scene shows Boethius on the left, a herald of modernity, performing a calculation with a pen in Arabic numerals, Pythagoras on the right, representing the ancient world, calculating with a line abacus, and in the middle Arithmetic represented by a young woman, who seems to have made her choice by looking towards Boethius. However, one may be surprised at the anachronism of this allegory, as Boethius, who died in 524, could certainly not have known Arabic numerals and Indian calculation.

Within this long history, it is appropriate to mention a singular character: Gerbert of Aurillac (born between 945 and 950, died in 1003), who was pope between 999 and 1003 under the name of Sylvester II (Otisk, 2022; Schärlig, 2022). He is credited with the invention of an instrument consisting of a calculation table and a set of horn tokens each marked with a number (Fig. 2.3). Since the tokens used are not all identical, this new instrument no longer strictly corresponds to the definition we gave at the beginning of the word "abacus"; it is nevertheless called "Gerbert's abacus" by extension of meaning. In the first version of the instrument, the numbers followed the *ghubar* script, which is to say that of the Western Arabic numerals in use in the Maghreb and Andalusia, which, through successive deformations, gave

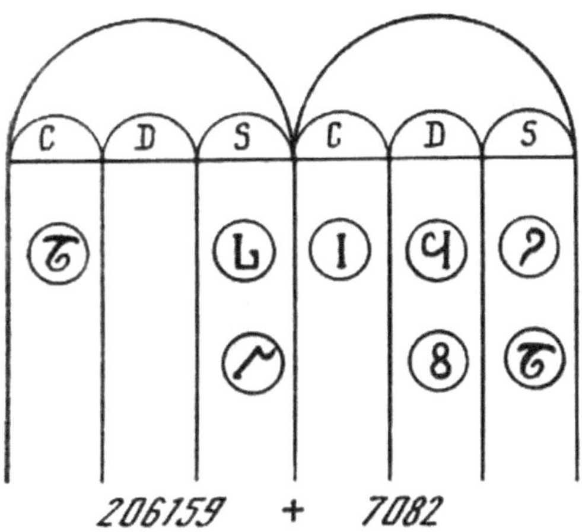

Fig. 2.3 Gerbert's abacus

us the familiar Arabic numerals (in the Eastern Arab countries, other numerals are used, closer to the original Indian numerals). To perform additions with Gerbert's abacus, it is therefore necessary to have nine piles of tokens and to know one's addition tables, which was not necessary with the old abacus. We can see that with this instrument, we are very close to our positional numeration.

To understand where this new artefact comes from, it should be known that Gerbert received part of his education in the Catalan abbey of Ripoll, a place of study equipped with a scriptorium where monks copied Latin translations of Arabic works on, among other things, astronomy, geometry and arithmetic. The county of Barcelona had become Christian again in 801 after being under administration in the name of Islam for a century and a half, making it a privileged place of intercultural encounter. The most probable hypothesis is, therefore, that Gerbert became aware of Indian calculation and Arabic numerals in Ripoll, and that he made a synthesis between this new knowledge and the abacus with tokens he was used to, to design his new abacus.

Why stop halfway instead of fully adopting Indian calculation? After all, instead of placing a token marked with the number 8 on the abacus, one could simply write the number 8 on any writing medium after having drawn columns, or even no columns at all if one takes care to respect the correct alignment of the numbers. On the one hand, as already mentioned, few people knew how to write; everyone was familiar with token abaci, and it was not advisable to disrupt habits too much. On the other hand, Gerbert was probably afraid of being accused of heresy if he openly adopted Arabic techniques. There are indeed later versions of his abacus in Greek alphabetical numbers, and even in Roman numerals, versions easier to accept within Christendom. After his death, Gerbert will still be accused of witchcraft, but for other reasons: he will be suspected of having made a pact with the devil among the Arabs of *Andalus* and of having become pope by sorcery.

Calculating with a Historical Abacus?

As we have seen, the transition from abaci to written calculation was a long and difficult process. Currently, we are experiencing another transition of the same magnitude: the transition from written calculation to electronic calculation. Once school and college years are over, practically no one regularly performs operations by writing, especially since calculators have recently been integrated into smartphones and tablets, objects that are constantly within reach of new generations.

Our working hypothesis is that it is relevant from a didactic point of view for children to relive the history of calculation in fast-forward: the transition from the token abacus to written Indian calculation, then the transition from the latter to electronic calculation. We are interested here in the first stage of this process. For the youngest, the token abacus can be an effective aid for learning decimal positional numeration and operational techniques with small whole numbers. Subsequently, it can be used to consolidate achievements, for possible remediation for struggling pupils, and to extend operational techniques to large whole numbers and decimal numbers. Teachers will find here a progression and some session diagrams to go in this direction.

Scientific and Pedagogical Aspects

The Difficulties of Pupils

Surveys and evaluations conducted in France (Ministère de l'Éducation nationale, 2015; CNESCO, 2017) revealed that 42.4% of 10–11 year old pupils have a fragile mastery of mathematics, particularly in numeration and calculation: one in four pupils would not know how to write a whole number greater than 10,000 in numbers; moreover, 20% of pupils do not succeed in a subtraction with carryover. These results were confirmed by the TIMMS surveys in 2015, 2019 and 2023. To combat these disappointing results, it is currently recommended to develop the manipulation of objects throughout primary school, and not just in kindergarten.

Internationally, various pedagogical and didactic research has been conducted on the use of the Chinese abacus as a tactile and visual support to facilitate the learning of numeration. Tsiapou and Nikolantonakis (2016), then Popotis and Nicolantonakis (2019) have shown how the Chinese abacus allows pupils to progress in the acquisition of the sense of the decimal numeration system, addition and subtraction. For her part, Poisard (2016, 2006, 2017) relied on physical, virtual and drawn Chinese abaci on a sheet to provoke register changes likely to promote the understanding of these same numerical notions. From the point of view of the history of mathematics education, Freiman and Volkov (2019) have also studied the history of the use of different forms of abaci, particularly the Russian abacus, in mathematics classes.

While bead abaci inspire numerous studies, the same cannot be said for token abaci, which seem less known and less utilised by teachers and teacher trainers. However, our experiences at various levels have convinced us that it is also an excellent hands-on tool to address pupils' difficulties.

A Remediation Approach: The Token Abacus

The token abacus has the advantage of being easily usable in a classroom, as it can be made at a low cost and can be easily stored in a notebook or in the desk compartment. It can be used in the context of numeration and calculation, preferably throughout the year. In lower classes (6–8 years), it seems preferable to stick to an abacus without quinaries organised in columns, so that there is a clear and direct correspondence between the physical representation of a number on the abacus and its notation in the units, tens, hundreds, etc. For older pupils (9–11 years), we have chosen an abacus with quinaries organised in rows. Historically, this was the most used configuration, as it corresponded well to the writing of numbers in Roman numerals (I, V, X, L, C, D, etc.). From a didactic point of view, it introduces a change of scenery that allows for a stimulating and non-repetitive review of previous years' achievements. This change of scenery is comparable to that of the Mayans' "dot-bar" number writing, exploited in the first chapter of this book.

Representing a Number on the Token Abacus

The token abacus we use consists of horizontal lines representing the different orders: units, tens, hundreds, thousands, tens of thousands, etc., and a vertical line, a separation mark that allows two numbers to be written face to face.

To represent a number, it is enough to place, for each order, as many tokens as the value of the corresponding digit. In order to reduce the number of tokens to handle, five tokens on a line are replaced by a single quinary token positioned anywhere between this line and the line immediately above. When there are no more than four tokens on a line and no more than one token between two lines, we speak of a reduced representation. Thus, for each order, the reduced representations of the ten "digits" on the abacus are shown in Fig. 2.4.

As an example, let's explain how to represent the number 158,074 on the abacus (Fig. 2.5). We place four tokens representing the units on the lowest line. We then represent the seven tens ($7 = 5 + 2$) by two unary tokens on the tens line and a quinary token just above. The zero marking the absence of hundreds, there is nothing to place on the hundreds line. The eight thousands ($8 = 5 + 3$) are represented by three unary tokens on the thousands line and a quinary token just above. All that remains is to place a quinary token above the tens of thousands line and a unary token on the hundreds of thousands line.

Once the representation of numbers is mastered, the two columns allow for exercises comparing two numbers. The larger number is the one that occupies a higher level, or uses more tokens at the highest level.

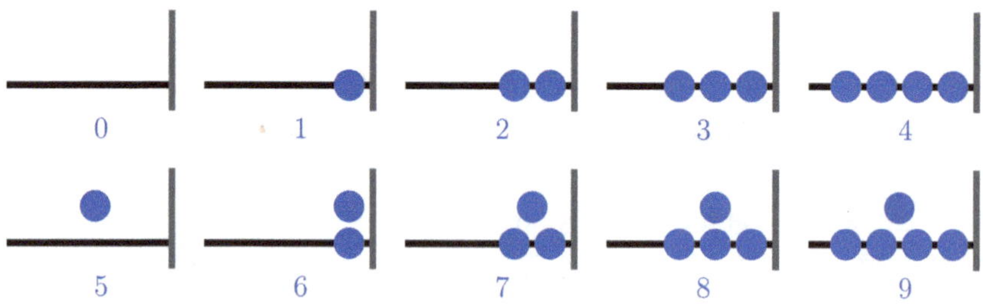

Fig. 2.4 The reduced representation of the ten "digits" on the abacus

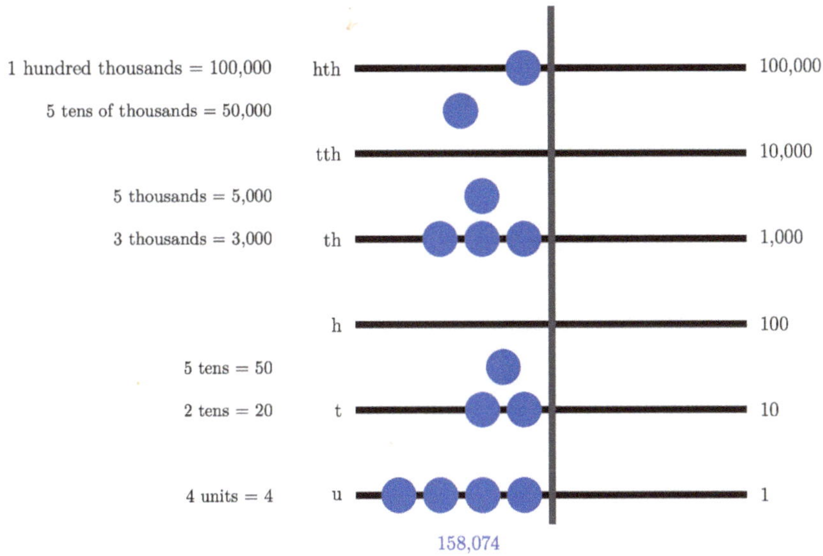

Fig. 2.5 Example of how to represent a number on the abacus

Performing an Addition on the Token Abacus

To add two numbers, we first place them side by side, then simply move the tokens from the right column to the left column. This reinforces the meaning of addition by reminding through manipulation that it involves counting the union of two collections of objects. We finish by making reductions according to the following rules:

- five unary tokens of the same order are exchanged for a quinary token of the same order (therefore in the space just above);
- two quinary tokens of the same order are exchanged for a unary token of the higher order (therefore on the line just above).

Example: calculating the sum of 11,259 and 7,031 (Fig. 2.6).

- Step 1: "place" the two numbers on each side of the abacus
- Step 2: "move" the tokens from left to right, materialising the addition
- Step 3: "exchange" five unaries for a quinary
- Step 4: "exchange" two quinaries for a unary
- Step 5: "read" the sum, here 18,290

Performing a Subtraction on the Token Abacus

On the token abacus, subtracting is simply placing the tokens representing the larger number, then removing the tokens representing the smaller one. By manipulation,

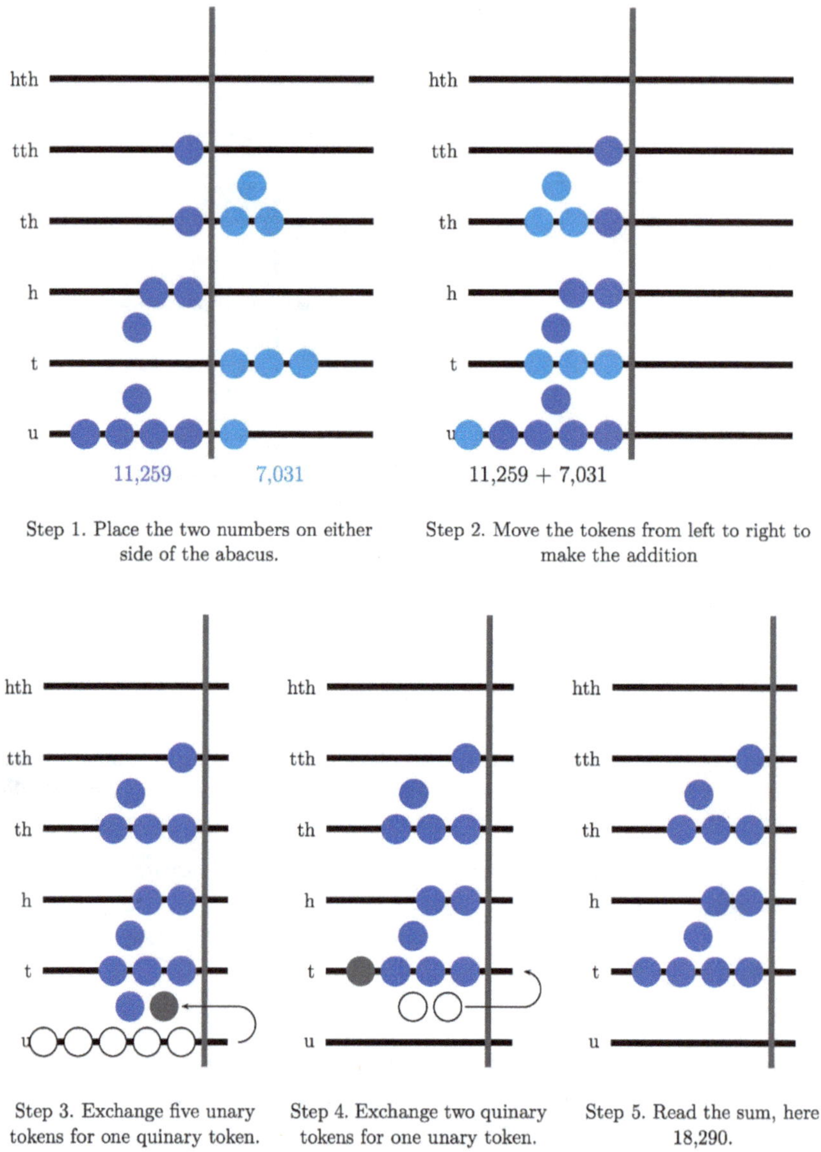

Step 1. Place the two numbers on either side of the abacus.

Step 2. Move the tokens from left to right to make the addition

Step 3. Exchange five unary tokens for one quinary token.

Step 4. Exchange two quinary tokens for one unary token.

Step 5. Read the sum, here 18,290.

Fig. 2.6 Example of calculating the sum of two numbers on the abacus

we realise that it is the inverse operation of addition: instead of adding a collection of objects to another collection of objects, it is about extracting a collection from another.

More precisely, to subtract one number from another, we place the larger number on the left and the smaller one on the right: we could, of course, subtract the smaller number from the larger number without placing the smaller one, but

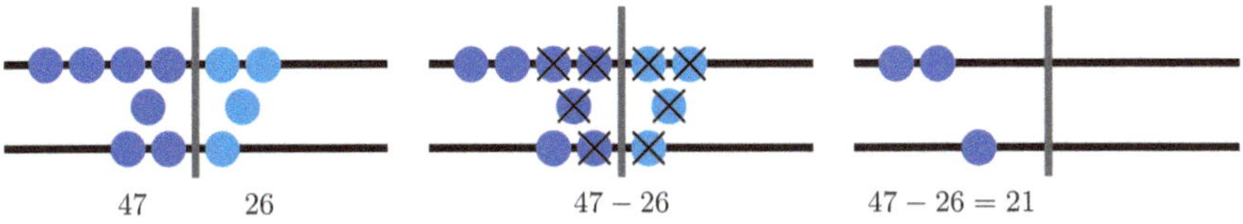

Fig. 2.7 Subtraction on the abacus without exchange

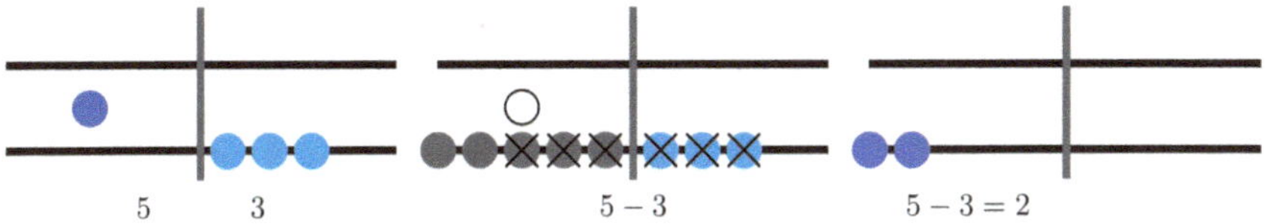

Fig. 2.8 Subtraction on the abacus with exchange of a quinary

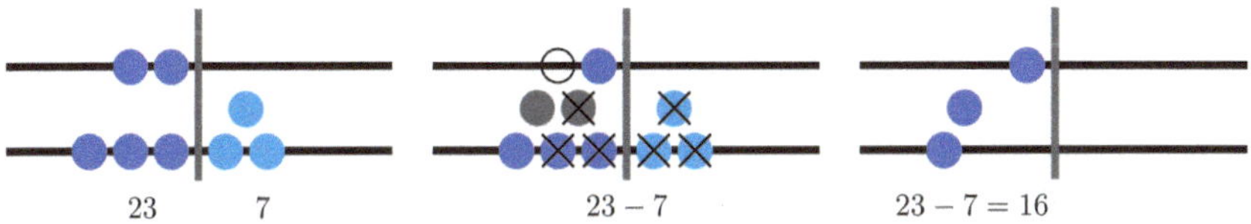

Fig. 2.9 Subtraction on the abacus with exchange of a unary

by keeping it in mind and subtracting order by order. However, placing both numbers avoids cognitive overload due to the large number of actions to be performed simultaneously (remembering the number, removing tokens, grouping-exchanging…). Several cases may arise:

- when, for each order or sub-order (the space of the quinaries), there are more tokens on the left than on the right, we remove as many tokens on the right as on the left until there are none left on the right (Fig. 2.7);
- when, for an order or sub-order, there are more tokens on the right than on the left, we must proceed to exchange a token of higher value for tokens of lower value (this operation, inverse of the reduction, is called amplification):
  - if we cannot remove unaries, it is enough to exchange a higher quinary, which then transforms into five unaries (this is the first type of amplification). After this exchange, all that remains is to remove the tokens of the same value on the right and on the left (Fig. 2.8);
  - if we cannot remove a quinary token, it is enough to exchange a unary of the higher order, which then transforms into two quinaries (this is the second type of amplification). In the same way as previously, all that remains is to remove the tokens of the same value on the right and on the left (Fig. 2.9).

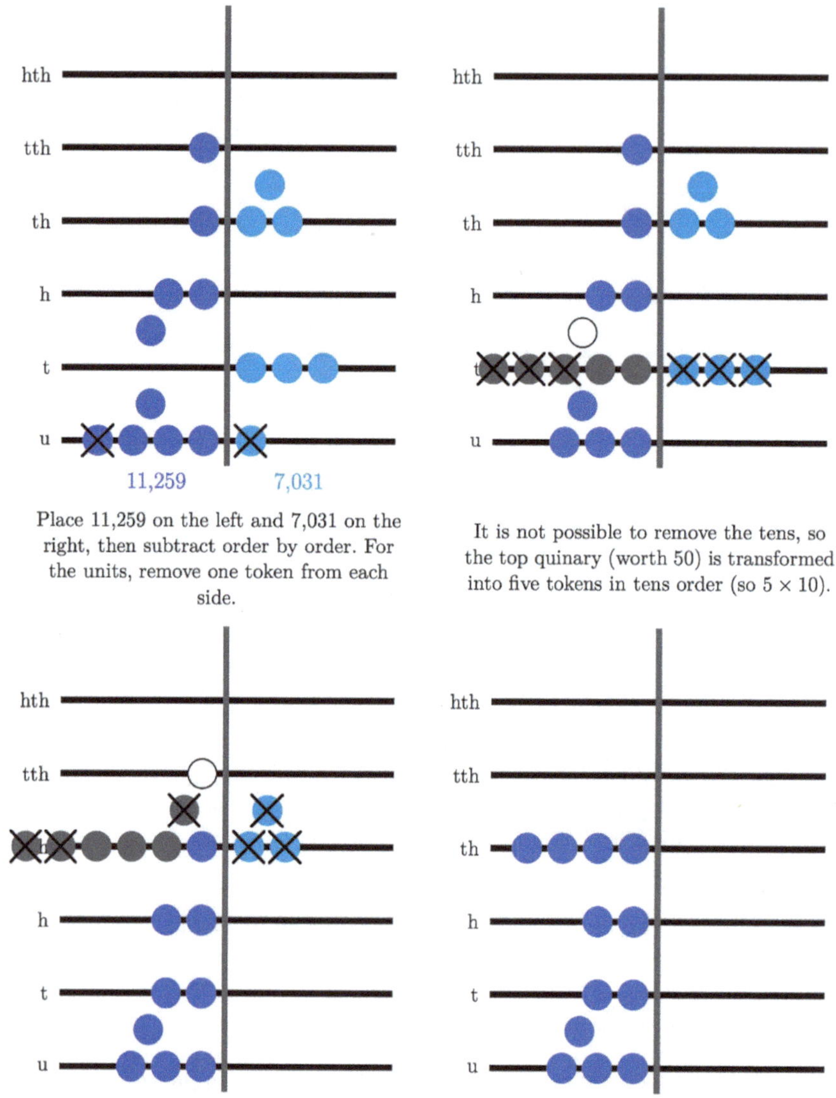

Place 11,259 on the left and 7,031 on the right, then subtract order by order. For the units, remove one token from each side.

It is not possible to remove the tens, so the top quinary (worth 50) is transformed into five tokens in tens order (so 5×10).

It is not possible to remove the thousands, and there are no quinary thousands to amplify, so we transform the tens of thousands token into two quinary thousands tokens, then one quinary token into five unary tokens.

Finally, the number obtained is read out : $11,259 - 7,031 = 4,228$.

Fig. 2.10 Example of how calculating the difference of two numbers on the abacus

Example: subtract 7,031 from 11,259 (Fig. 2.10).

We place 11,259 on the left and 7,031 on the right, then proceed to subtractions order by order. For the units, we remove a token from each side.

It is not possible to remove the tens, so we transform the quinary above (which is worth 50) into five tokens of the order of the tens (so 5×10).

It is not possible to remove the thousands and there is no thousand quinary to amplify, so we transform the token of the order of tens of thousands into two tokens of thousand quinary, then one token quinary into five tokens of units.

Finally, we read the number obtained: $11{,}259 - 7{,}031 = 4{,}228$.

Implementation in the Classroom

The Material

For optimum results, the use of the token abacus should be done as a common thread throughout the year, as a tool for representation of numbers, comparison, calculation, verification and remediation. For this, there is educational material available in the market, such as the "Kit calculus" by Michel Mouyssinat (2015, 2018), which we have experimented with and which we have largely inspired to create personal tools afterwards (Daval, 2014, 2015, 2017). Indeed, the necessary material is easily achievable by a few means; here are some ideas for choosing tools (also consult the Appendix):

- **choice of the token abacus**: depending on the needs of the class and the objectives to be achieved, one must start by asking the question of the number of orders that one wishes to make appear on the abacus. A simple text or image processing allows to build quite quickly an abacus made of horizontal lines and a vertical separation (you can also simply draw it by hand, or even have it drawn by each pupil). Next to each line, it is possible to have a coding to help pupils to find their way: the initial of the order, its representation in letters or in numbers (for example, "100", "h" or "hundreds");
- **creation of token abaci**: once drawn, it is enough to print them in A4 format, then to laminate them in order to guarantee a durable use. For a tool usable "on the fly", it can also be interesting to create a desk pad including a part of token abacus, therefore always visible and within the pupil's reach;
- **choice of tokens**: again, the choice is vast and will depend on the size of the lines and the spaces between the lines. It must be possible to put at least five tokens on each line and two tokens between two consecutive lines. You can use game tokens, Picbille, dry beans, Cape peas or any object easy to obtain in number and that does not roll. An alternative is the use of an erasable marker, whose advantage is to create tokens that do not move once placed on the abacus. However, this loses the tactile manipulation of the tokens and the grouping-exchanges they induce;

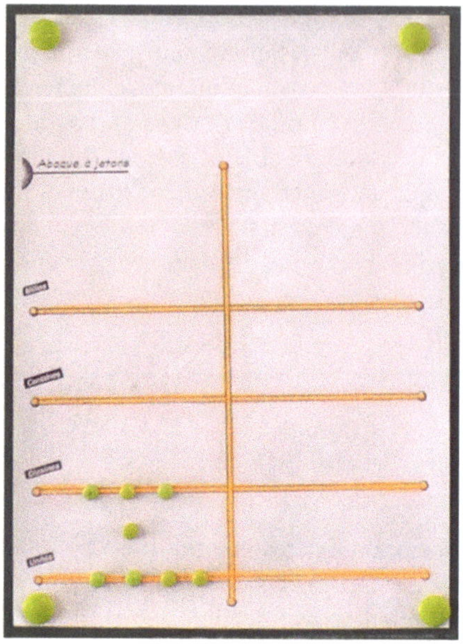

Fig. 2.11 A magnetic board abacus, with magnets as tokens (Photo: Nathalie Daval)

- **creation of a board abacus**: the same abacus as that of the pupils can be printed in A3 format at least and fixed to the board with magnets, for example. Other magnets will act as tokens (Fig. 2.11). There is also erasable and adhesive surface paper to stick on a corner of the board or on the classroom wall, to use with an erasable marker (Fig. 2.12). You can finally use a projector to project an abacus on a whiteboard and write on it;
- **"paper-pencil" sheets**: any work of manipulation on the abacus can be associated with a written work of coding and decoding. Just use the chosen abacus model and reduce it in order to offer pupils sheets where they can either represent numbers on the abacus, or read numbers that they will have previously represented on the abacus.

The ideal is for each pupil to have their own abacus so that they can manipulate it whenever they wish, even if some activities are better conducted in pairs, using a single abacus for two pupils.

Abacus and Numeration

A sufficient period must be devoted to the representation of numbers and their comparison on the abacus. Until this stage is fully understood by the pupils, it is not reasonable to use the abacus as a calculation tool in parallel with expert techniques with the aim of explaining how these work.

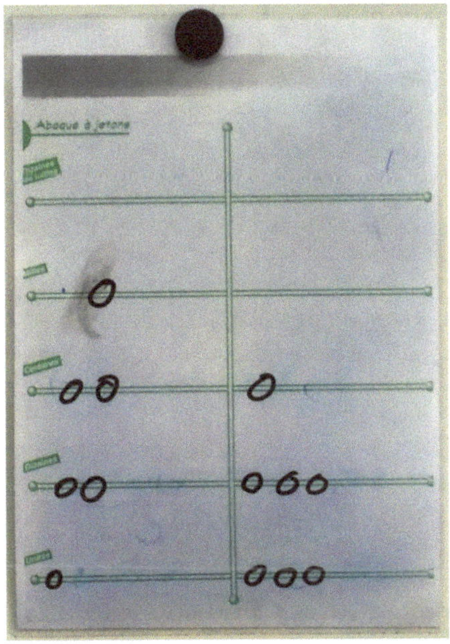

Fig. 2.12 A magnetic board abacus, with tokens drawn with erasable marker (Photo: Nathalie Daval)

An Example of Progression

Here is a framework for using the token abacus in the context of a learning or remediation in numeration. Initially, for all new steps, it is preferable to let the pupils work in pairs; it is only when understanding is sufficient that pupils can use their personal abacus. The durations indicated are average indicative durations, to be adapted according to the age and level of the pupils.

1. **Step 1: discovery of the abacus** (*a session of 45 minutes*)
   This is about presenting the abacus, with a historical opening on the evolution of our numeration and the calculation tools that preceded digital tools. The teacher can then leave a free period of manipulation and questioning: what can the lines be used for? How to place the tokens?... To finish, some pupils propose a use and exchanges are established within the class on this subject.
2. **Step 2: writing the first numbers** (*a session of 45 minutes*)
   For this step, one can start by not introducing the quinary system right away and ask the pupils how to write all the numbers from 1 to 10.
   Once the system is understood, one can ask the pupils which numbers it is possible to write with a single token (depending on the class level, we will go up to 10,000 or up to a billion). On this occasion, we can make the link with our positional numeration: indeed, a token outside the abacus has no value; it is "a token"; it takes all its numerical power when it is placed on the abacus, depending on its position.
   Finally, we can repeat the same work with two tokens.

3. **Step 3: introduction of the quinary** (*a session of 45 minutes*)
The writing of numbers greater than five for a particular order is an opportunity to raise a spatial difficulty: too many tokens do not aid either reading or manipulation. Hence the need to find a more concise way of writing numbers by introducing a token worth five tokens on its own: it is called the quinary token or simply the quinary. This important step is far from easy to bring about and it will certainly be necessary to give a boost to the pupils. The difference in roles between a quinary token and a token that represents the ten (which defines the base) must be carefully highlighted by the teacher.
The work of step 2 can then be repeated considering that a token can be "unary" or "quinary". To convince the pupils of the usefulness of the quinary token, particular attention is paid to the fact that it is easy to count a quantity less than five by simple visual perception, whereas it is not the case for larger quantities. This also leads to working on the additive decompositions of numbers less than 10 based on 5, very useful for mental calculation.
4. **Step 4: Writing numbers up to 100** (*one to two session(s) of 45 minutes*)
From this step, we can represent numbers up to 100. It is also important to propose the reverse work: represent a number on the abacus and have the pupils read it. We can, for this purpose, operate in a whole class thanks to the abacus on the board, for example, with writing on the slate, or use worksheets to complete.
5. **Step 5: Creation of a summary poster** (*one 45-minute session*)
Once the system is understood for numbers up to 100, we propose to the pupils to create a summary poster. This work can be done by small groups. We can create the class poster from the pupils' work, then hang it on the classroom wall.
6. **Next steps: Reading and writing larger numbers** (*one to two 45-minute learning session(s), then ongoing improvement as part of routines, for example*)
Once the first five steps are understood, we increase the size of the numbers: up to 1,000, then 10,000... up to billion.
Abaci allow the manipulation of real objects that are the tokens, but we must not forget to make the parallel with our more conceptual numeration (with the classic numeration table, for example), the goal being, *ultimately*, a better understanding of our decimal positional numeration system.

Difficulties and Remediation Tips

The Quinary It is not obvious for pupils to find the existence of the quinary on their own and several steps are necessary to bring them to assimilate its position and value (Figs. 2.13 and 2.14); the acquisition of the quinary must therefore be done gradually through multiple manipulations, the pupils "forgetting" its use from one session to the next (unless they have already practiced Roman numeration).

A possible step would be to model (initially) this quinary token by a larger token or a different colour, in order to represent it differently from a unary, but this subterfuge must remain temporary or it could become an obstacle to learning. Indeed, it is essential that pupils understand that the value of a token depends only on

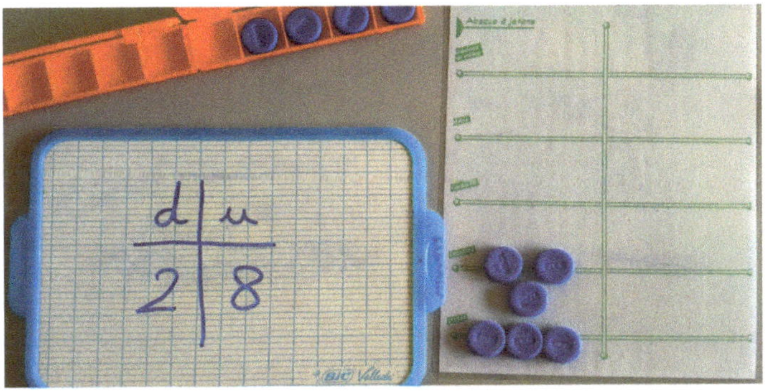

Fig. 2.13 This pupil has used the writing of the number in a numeration table, he has correctly represented the number 28 on his abacus, the quinary is well placed (Photo: Nathalie Daval)

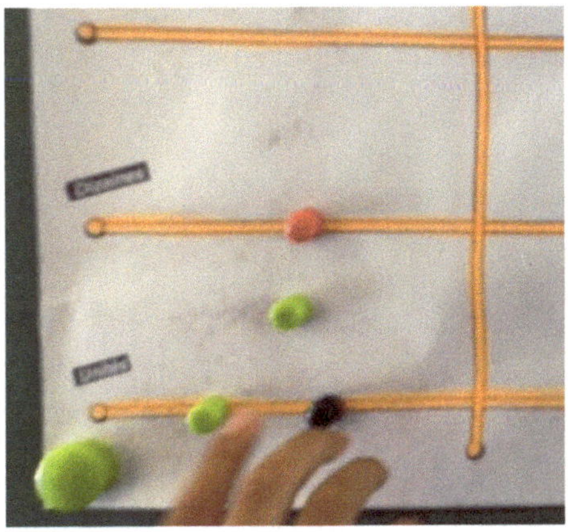

Fig. 2.14 Here, the pupil had to read the number proposed on the abacus, his answer is "31", he therefore did not perceive the role of the quinary token, nor the different lines of the abacus (Photo: Nathalie Daval)

the location of this token on the abacus to the exclusion of any other consideration, just as in written numeration, the value of a digit depends only on its position in the writing of a number.

Position of the Abacus Initially, the vertical position of the abacus can be a hindrance, since numbers are read vertically from top to bottom, whereas in everyday life, we read them horizontally from left to right. As mentioned above, it is a fact that this orientation of the abacus is part of the history of the instrument. However, if the teacher deems it preferable, a horizontal use of the abacus is entirely possible, as is indeed the case with bead frames or rod abaci.

Forgetting from One Session to the Next If the abacus is not used regularly, pupils tend to forget its characteristics, particularly in relation to the quinary system. That's why only frequent use will lead to automation. This can be part of a ritual

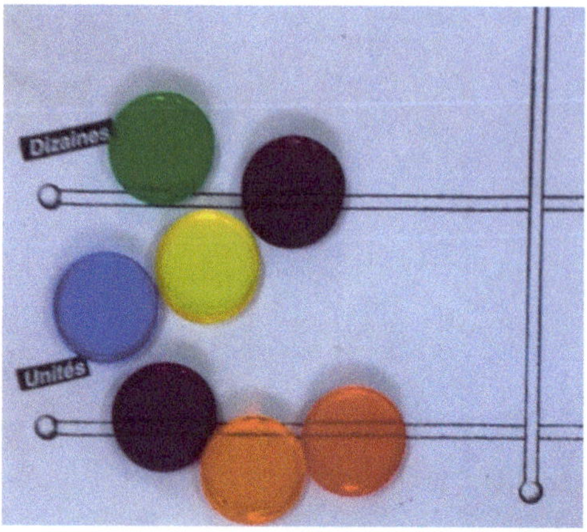

Fig. 2.15 Tokens placed "anyhow" on and off the lines, so that it is impossible to read the number represented on the tens line (Photo: Nathalie Daval)

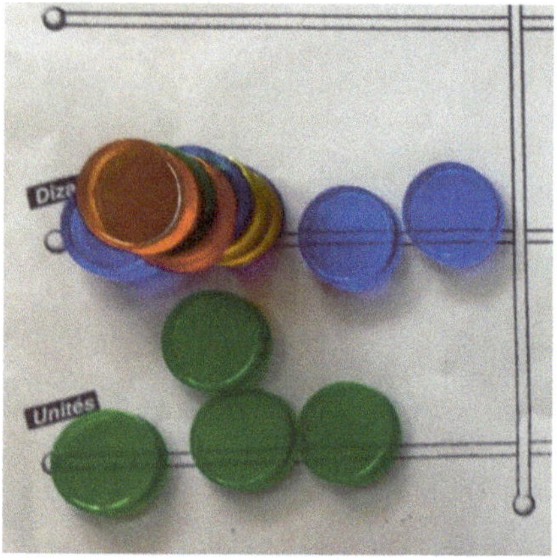

Fig. 2.16 Tokens placed "roughly" on the lines, but stacked on top of each other (Photo: Nathalie Daval)

moment (at the start of the day/half-day, at the start of the maths session, after break. . .): for example, the teacher suggests a number that the pupils must represent on their abacus, and conversely, he shows a representation on the abacus on the board, which the pupils must read and write on their slate.

Manipulation of Tokens The motor difficulty caused by the tokens used depends on their shape and size. Too thin or too small, they are difficult to pick up. Conversely, the larger they are, the more space they take up and tend to get mixed up. Apart from this consideration, the use of the abacus requires flawless organisation: it is a matter of not stacking the tokens and positioning them correctly on the lines and between the lines (Figs. 2.15 and 2.16).

Abacus and Addition

Addition with the abacus can be introduced by a problem situation leading to the establishment of an expert procedure directly resulting from the pupils' trial and error. Then, additions to the abacus are an opportunity to perfect the understanding of the operation set out in columns by making the link, at each stage, with this.

An Example of Progression

1. **Step 1: discovery of addition on the abacus** (*a 45-minute session*)
   A problem situation in the additive field (and more specifically with the objective of having to perform an addition) is an opportunity to discuss with the pupils the means of setting and performing an addition on the abacus. It is up to them to experiment and find solutions based on their knowledge, solutions that will be debated, refuted or validated in the class group.
2. **Step 2: represent two numbers on the abacus and compare them** (*a 45-minute session*)
   The comparison is made directly, by "correspondence order by order" starting with the highest order, that is to say for the token(s) located highest on the abacus. This first step aims to confront the pupil with two numbers placed on the abacus, to realise that tokens on the same line or the same space on each side of the median line have the same value. It also anchors the pupils' knowledge on the comparison of large numbers in their decimal writing and facilitates entry into calculation, as they will then have to add tokens of the same value.
3. **Step 3: add two numbers without grouping or exchanges** (*a 45-minute session*)
   Two numbers are proposed, for which no group-exchange is necessary. The pupils place the tokens on the same side, being careful not to change the order, and directly read the result of the operation.
4. **Step 4: Add two numbers with unaries-quinary exchanges** (*one to two session(s) of 45 minutes*)
   During this step, two numbers are proposed, for which in one or several of the orders, five tokens or more will appear after movement. A reduction must then be carried out, that is an exchange of five unary tokens for a quinary token (of the same order). Particular attention is paid to the way this reduction is carried out: the tokens are physically taken from a line and replaced by a single token (other than the five tokens to be exchanged) on the space just above.
5. **Step 5: Add two numbers with quinaries-unary exchanges** (*one to two session(s) of 45 minutes*)
   The principle is the same as before, but here, exchanges of two quinaries for a unary of the higher order are carried out.
6. **Step 6: Add any two numbers** (*one to two session(s) of 45 minutes of learning, then use in continuous thread*)

Once the first five steps are acquired, it is then possible to carry out any addition on the abacus, with increasingly large numbers depending on the progress of the class in the numeration of whole numbers.

7. **Next steps**:
   use of the abacus as a calculation tool during traditional calculations and during problem-solving.

Difficulties and Remediation Paths

The main difficulties come from the group-exchanges and the movements they imply. These group-exchanges correspond to the carries in a traditional column operation. The difference is that there are two types of group-exchanges in an abacus calculation, while there is only one in the written calculation, the latter not calling on the notion of quinary.

Group-Exchanges

There are two types of exchanges: the "five unary tokens for one quinary token of the same order" exchanges and the "two quinary tokens for one unary token of the higher order" exchanges. The first exchange mentioned generally poses fewer difficulties, as it involves adding units to create a grouping by five (which is somewhat akin to saying that there are five fingers in a hand).

On the other hand, the second exchange is more difficult to grasp, as the quinary tokens are not differentiated, and pupils often "forget" their value to deduce that a quinary plus a quinary makes two unary tokens. In this case, the previous suggestion to represent the quinary differently could be a temporary option.

The Movement of Tokens

A first mistake is manipulative: pupils move tokens onto an adjacent line, or they have placed too many on the line and some end up in the space just above or below, suddenly changing their value.

Sometimes, pupils make hasty exchanges (especially when they start to get a little more comfortable with the token abacus): for example, when exchanging five unary tokens for a quinary, they remove four tokens and move the last one to the space just above. If the result is correct and the shortcut is acceptable for an expert user, the concept underlying the manipulation may be questioned: it is no longer an exchange, but a movement of tokens, and the usefulness of the abacus becomes less effective.

Another way of doing it, still for the same case, but this time incorrect, is to take one of the unary tokens already placed on the abacus and move it to the quinary position while forgetting to remove the four remaining unary tokens on the abacus.

These two procedures are not necessarily attributable only to pupils, as they may result from learning by imitation: the teacher or the pupil who makes exchanges on

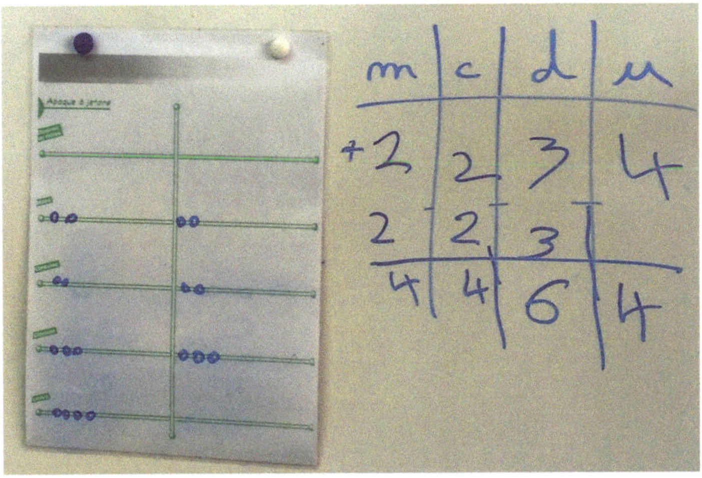

Fig. 2.17 Incorrect positioning of the number 223, aligned with the thousands digit of the number 2,234 (Photo: Nathalie Daval)

the board must therefore be vigilant about the need to adopt correct gestures to avoid creating additional obstacles.

The Placement of the Two Numbers on the Abacus
The pupil may align the smaller number not on the order of the units, but on the highest order of the larger number (Fig. 2.17). Just like the position of the different digits of a number in a column addition, one must be vigilant about the positioning of the two numbers on the abacus and have the represented numbers read before starting the addition (Fig. 2.18).

Ways to Go Further

Following these two examples (the abacus for the decimal positional numeration representation of natural numbers and the abacus for addition), the work can be expanded to decimal numbers and other operations.

Decimal Numbers

To use the abacus to work with decimal numbers, simply create an abacus including the orders of tenths and hundredths at the beginning, then thousandths and ten-thousandths for more experienced pupils. The process is then identical to that implemented for whole numbers. If pupils have gotten used to using the abacus from

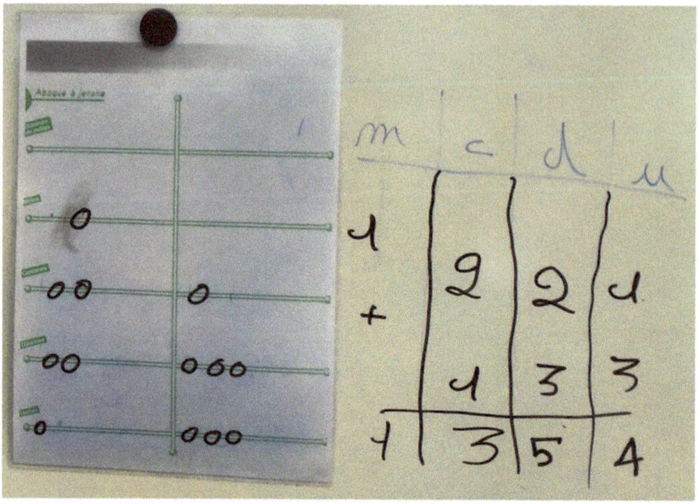

Fig. 2.18 Correct positioning and operation (Photo: Nathalie Daval)

the beginning of the year and throughout, this step should not be an insurmountable obstacle and should allow for a better understanding of how to write, read, and represent decimal numbers. A definite advantage is writing such a number on the abacus: indeed, very often, pupils consider a decimal number as "two whole numbers separated by a comma". In the case of the abacus, it is a single number.

The sum of two decimal numbers can also be performed on the abacus: for example, the operation $2.13 + 3.4$ often gives the result 5.17 since the pupils add the whole parts separately ($2 + 3 = 5$), then the decimal parts ($13 + 4 = 17$) without worrying about the link that unites them. On the abacus, and provided that the numbers are well represented, this error should be avoided.

Other Operations

The four basic operations can be performed on the abacus. This tool will provide real added value for subtraction, especially in the case where the pupil is used to performing subtraction "by borrowing", since the method on the abacus is almost identical: to remove a number from a smaller number for a certain order, it is necessary to "break" the higher order in order to redistribute it to the current order.

The setting of the multiplication requires a good mastery of addition, since the result comes from a series of multiplicative and additive calculations: we decompose the multiplier in canonical form, then we perform each of the multiplications of the orders of the multiplier by the multiplicand, results that will then have to be added.

Finally, division requires an additional skill. It is the most difficult basic operation to perform on the abacus, but it can allow some advanced pupils to take on increasingly complex challenges.

Acknowledgments This chapter was written by Nathalie Daval and Dominique Tournès on behalf of the research group on abacuses and bead frames of the IREMI of Reunion (France). It owes much to Luc Tiennot, trainer at the INSPE of Reunion, and to the five primary school teachers who experimented with the use of the token abacus in their classes: Catherine Dugain and Marie-Jeanne Guillotin (2014–2015), Amandine Grondin and Sylvie Grondin (2015–2017) and Myriam Trémoulu (2016–2017). It follows on from the work carried out in partnership between the IREMI of Reunion and the INSPE of Brittany (team of Caroline Poisard) during the years 2013 to 2016, work which led to the publication of special issue 51 of *MathémaTICE*. We refer to this issue for a didactic study of bead abacuses, which we do not deal with here, as well as for a more complete version of the history of abacuses.

Appendix 1: Concepts and Skills Addressed in This Chapter

Covered Concepts

- Use and represent large whole numbers
  - Compose, decompose large whole numbers, using groupings by thousands (units of numeration: simple units, tens, hundreds, thousands, millions, billions and their relationships).
  - Understand and apply the rules of numeration to large numbers (up to 12 digits).
  - Compare, order, frame large whole numbers.
- Calculate with whole numbers
  - Memorise numerical facts and basic calculation procedures. Develop or choose calculation strategies orally and in writing.
  - Written calculation: implement a written calculation algorithm for addition, subtraction.

Developed Skills

- **Search and research skills**: Extract, organise information, manipulate, experiment from a new support: the token abacus.
- **Reasoning**: Perform calculations requiring the construction of a process that combines reasoning steps (group to better count, group-exchange to calculate).
- **Computational skills**: Use a new operational technique on the abacus in connection with traditional techniques.
- **Mathematical communication skills**: Use appropriate vocabulary to explain the use of the abacus, whether at the level of numeration or calculation, explain one's approach and argue in exchanges.
- **Mathematical representation ability**: Use the abacus to represent a number, produce different representations of this number; recognise an additive or subtractive situation, represent it on the abacus.

Appendix 2: Adjustable Abaci

Examples of printable abaci: some possible configurations depending on the objective to be achieved (pdf file available at https://www.univ-irem.fr/chapitre-2-de-l-abaque-a-jetons-au-calcul-pose) (Figs. 2.19, 2.20, 2.21, and 2.22).

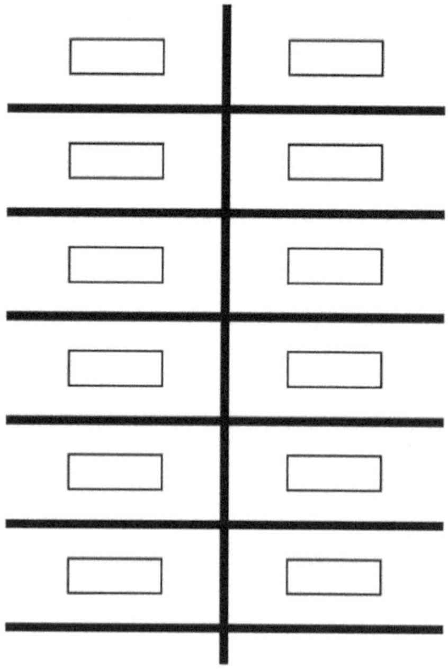

Fig. 2.19 Blank abacus with 6 lines, frame for quinaries

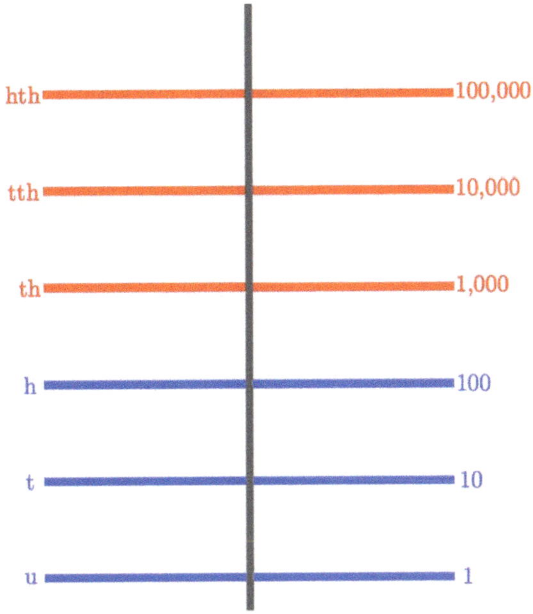

Fig. 2.20 Colour-coded abacus according to classes, letter and number indications, representation from 0 to 999,999

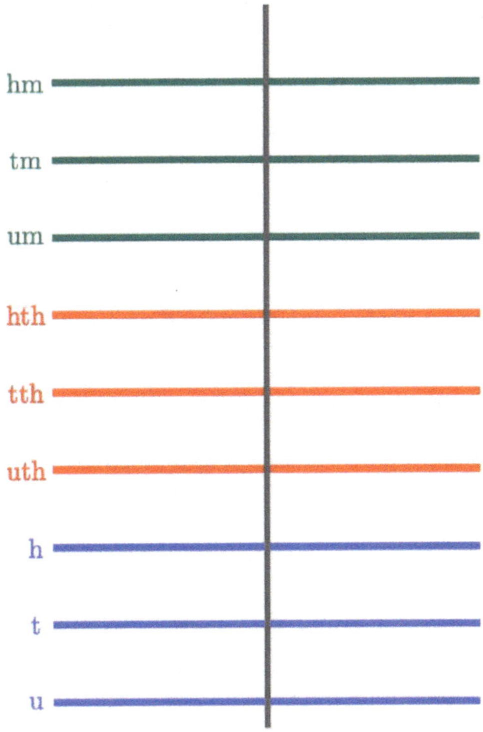

Fig. 2.21 Colour-coded abacus according to classes, letter indications, representation from 0 to 999,999,999

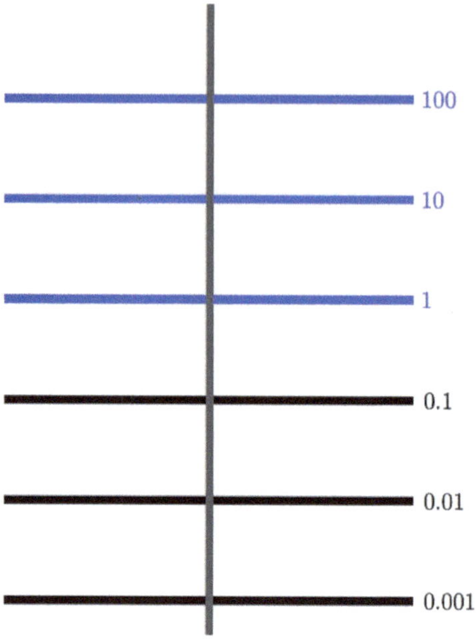

Fig. 2.22 Decimal abacus colour-coded according to classes, number indications, representation from 0.001 to 999.999

References

CNESCO. (2017). Nombres et calcul au primaire. http://www.cnesco.fr/wp-content/uploads/2017/02/FICHE_Nombre-et-calcul-au-primaire.pdf

Daval, N. (2014). Affiche de l'abaque à jetons. IREMI de la Réunion. https://www.univ-irem.fr/chapitre-2-de-l-abaque-a-jetons-au-calcul-pose

Daval, N. (2015). Tutoriel d'utilisation de l'abaque à jetons pour la numération. IREMI de la Réunion. https://www.youtube.com/watch?v=H7_2qI2Aa_8&index=6&list=PLO-BeM-MVUMioGTNRhs0tiIZ45W1iDbRT

Daval, N. (2017). Tutoriel d'utilisation de l'abaque à jetons pour l'addition. IREMI de la Réunion. https://www.youtube.com/watch?v=6SrCkP67wRc&list=PLO-BeM-MVUMioGTNRhs0tiIZ45W1iDbRT&index=2

Dehaene, S. (1617). *L'arithmethique. Departie en trois livres. Ensemble un petit discours des changes. Avec l'art de calculer aux getons*. Chez la veuve Jean Regnoul.

Dehaene, S. (1997). *The number sense. How the mind creates mathematics*. Oxford University Press.

Evans, G. R. (1977). From abacus to algorism: Theory and practice in medieval arithmetic. *The British Journal for the History of Science*, *10*(2), 114–131.

Freiman, V., & Volkov, A. (2019). Pedagogical value of the Russian abacus and its use in teaching and learning arithmetic in the 19th–early 20th century. In K. Bjarnadóttir, F. Furinghetti, K. Jenneke, J. Prytz, S. Gert, & H. J. Smid (Eds.), *"Dig where you stand" 5. Proceedings of the Fifth International Conference on the History of Mathematics Education* (pp. 161–176). Utrecht University, Freudenthal Institute.

Leibniz, G. W. (1765). *Œuvres philosophiques latines et françoises*. Schreuder.

Ministère de l'Éducation nationale. (2015). *CEDRE 2014 – Mathématiques en fin d'école primaire: les élèves qui arrivent au collège ont des niveaux très hétérogènes*. Ministère de l'Éducation nationale. Note d'information 18 de la DEEP.

Mouyssinat, M. (2015). Présentation du Kit Calculus. IREMI de la Réunion. https://www.univ-irem.fr/chapitre-2-de-l-abaque-a-jetons-au-calcul-pose

Mouyssinat, M. (2018). Le calcul aux jetons & Calculation with jetons. Presentation at ESU8, Oslo. https://www.univ-irem.fr/chapitre-2-de-l-abaque-a-jetons-au-calcul-pose

Moyon, M. (2016). *Fibonacci. Extraits du Livre du calcul (Liber Abaci).* ACL-Les Éditions du Kangourou.

Otisk, M. (2022). *Arithmetic in the thought of Gerbert of Aurillac.* Peter Lang.

Poisard, C. (2006). The notion of carried-number, between the history of calculating instruments and arithmetic. In P. Groontenboer, R. Zevenbergen, & M. Chinnapan (Eds.), *Identities, Cultures and Learning Spaces, Proceedings of the 29th Annual Conference of the Mathematics Education Research Group of Australasia*, (Vol. 2, pp. 416–423). MERGA.

Poisard, C. (2016). Les ressources virtuelles et matérielles en mathématiques : des instruments pour travailler en classe sur le nombre, la numération et le calcul. *MathémaTICE, 51.* http://revue.sesamath.net/spip.php?rubrique135

Poisard, C. (2017). Introducing an old calculating instrument in a new technologies environment: A praxeological analysis of students' tasks using different registers. *Re SMITECH Review of Science, Mathematics and ICT Education, 11*(2), 47–67.

Popotis, A., & Nikolantonakis, K. (2019). The contribution of the Chinese abacus to the development of the number sense. In E. Barbin, U. T. Jankvist, T. Hoff Kjeldsen, B. Smestad, & C. Tzanakis (Eds.), *Proceedings of the Eighth European Summer University on History and Epistemology in Mathematics Education (ESU8)* (pp. 268–277). Oslo Metropolitan University.

Pullan, J. M. (1969). *The history of the abacus.* Frederick A. Praeger.

Schärlig, A. (2001). *Compter avec des cailloux. Le calcul élémentaire sur l'abaque chez les anciens Grecs.* Presses polytechniques et universitaires romandes.

Schärlig, A. (2006). *Compter du bout des doigts.* Presses polytechniques et universitaires romandes.

Schärlig, A. (2022). *Gerbert d'Aurillac, mathématicien et pape de l'an mil.* Presses polytechniques et universitaires romandes.

Smith, D. E. (1921). *Computing Jetons.* The American Numismatic Society.

Sugden, K. F. (1981). A history of the abacus. *The Accounting Historians Journal, 2*, 1–22.

Tournès, D. (2016). Perspectives historiques sur les abaques et bouliers. *MathémaTICE, 51.* http://revue.sesamath.net/spip.php?rubrique135

Tsiapou, V., & Nikolantonakis, K. (2016). The development of place value concepts and the notion of carried number among sixth grade students via the study of the Chinese abacus. *Menon: Journal of Educational Research*, 2nd thematic issue, 51–65.

Nathalie Daval is a professor of mathematics at secondary school level and at the University of Montpellier for the training of schoolteachers. She also has a PhD in the history of mathematics, and works mainly in the fields of graphical calculation and early instrumented calculation techniques.

Dominique Tournès is professor of mathematics and history of mathematics at the University of Reunion Island (France), in the Computer Science and Mathematics Laboratory. As a historian of mathematics, his research focuses on the calculation methods and instruments that preceded the computer in the period 1750-1950. For many years, he has been involved in using history to improve the training of mathematics teachers.

Chapter 3
The Mechanisation of Calculation

Frédérique Plantevin

The machine, by its arrangement, perfectly imitates arithmetic operations, and its movements seem to paint all the reasoning that must be done to arrive at the results. [Our translation]

Louis-Alexandre-Désiré Hoyau (1822, p. 360)

Abstract The work presented in this chapter focuses on numeration and the calculation of arithmetical operations through mechanical instruments that have partially automated them. The sequence is based on the careful observation of two pin wheels calculators filmed in *ad hoc* videos. The machines are a "Lightning calculator" adder and an Odhner type "Brunsviga 20" multiplier.

These machines being at the same time mathematical and technological objects by nature, the activities carried out with them logically achieve objectives in the two disciplines, and beyond, as it were, through the linking of both of them. More specifically, working with these machines is designed to lead to a concrete, in-depth understanding of their operating principle which in turn, sheds new light on the mathematical objects and concepts involved in the operations they perform, like numeration basis and orders, carry-over, multiplier, quotient, divisor and remainder and also associativity and distributivity.

After presenting the machines and their historical context, the teaching sequence is detailed. Implementation in class is then presented and pupil productions are exhibited, illustrating the way the above objectives are achieved. The chapter ends with paths to go further in mathematics, in technology and in the history of science and technology.

Keywords Numeration · Pin wheel calculator · Arithmetical operations · Technology and mathematics activity · Materiality of mathematical concepts

F. Plantevin (✉)
IREM de Brest, Université de Bretagne Occidentale, Brest, France
e-mail: frederique.plantevin@univ-brest.fr

M. Moyon et al. (eds.), *Teaching Mathematics Through Historically-Based Activities*, History of Mathematics Education,
https://doi.org/10.1007/978-3-032-16990-7_3

Introduction

The work presented in this chapter focuses on numeration and the calculation of addition, subtraction, multiplication and division of natural numbers, through mechanical instruments that have partially automated them. It is designed for pupils between 9 and 11 years old, but may also be of interest beyond. We start by presenting the machines used by situating them in their historical context. The mathematical objectives that can be pursued around these machines are then exposed. Due to the interdisciplinary nature of the objects studied and the approach followed to use them, the activities also achieve objectives in technology and possibly in history (see the section "Ways to Go Further"). Accompanying videos, showing the machines in action, were made for the sequences in class. They are made available to everyone (see Appendix 5 for access to online resources). Watching them can aid the reading of this chapter.

Historical Context

The mechanisation of the four arithmetic operations began at the very beginning of the seventeenth century with the independent inventions of Wilhelm Schickard and Blaise Pascal. Their machines both perform additions of whole numbers with an automatic carry-over. They are based on a new representation of whole numbers, by suitably meshed graduated wheels. Each order corresponds to a wheel with ten teeth (or ten rods), which turns the one of the next order—placed to its left—after a complete turn. Pascal's machine, later called the *Pascaline*, was the starting point for numerous attempts to improve the automatic calculation of addition and subtraction on the one hand, and on the other hand, to enable that of multiplication (Cargou et al., 2025) and division (Marguin, 1994; Bruderer, 2020). The machines we worked with represent a culmination in these two directions, almost three centuries later. They are a wheel adding machine called *Lightning calculator* (Fig. 3.1) built in Michigan (USA) around 1930 according to a 1926 patent and several Odhner type multipliers (Fig. 3.2) manufactured between 1900 and 1960 according to an original patent filed in Sweden in 1878, and in France in 1892 by Willigot T. Odhner (see Scherphuis, 2024, for both machines, for instance).

The *Lightning calculator* wheel adding machine is a compact and fairly ergonomic office machine, inexpensive, which was manufactured until 1960. Simple in appearance, its sophistication lies in the compactness of its mechanisms (which can be observed in the patent), enclosed in a half-centimetre thick casing (see the casing separated from its support in Fig. 3.1), and their reliability. Later models will also allow subtractions, which is not the case with this one. To perform an addition, the operator enters the digits of each operand using the stylus. To enter 40, the stylus is inserted into the second wheel from the right, opposite the graduation 4, and brought against the stop, opposite the 0 of the dial, by turning

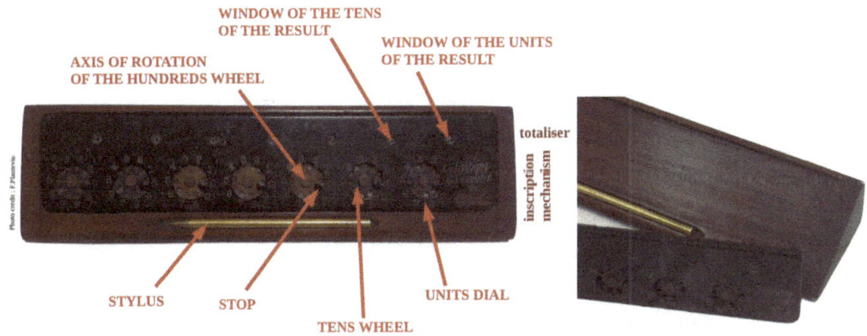

Fig. 3.1 Two views of the *Lightning calculator* adding machine (Photos: Frédérique Plantevin)

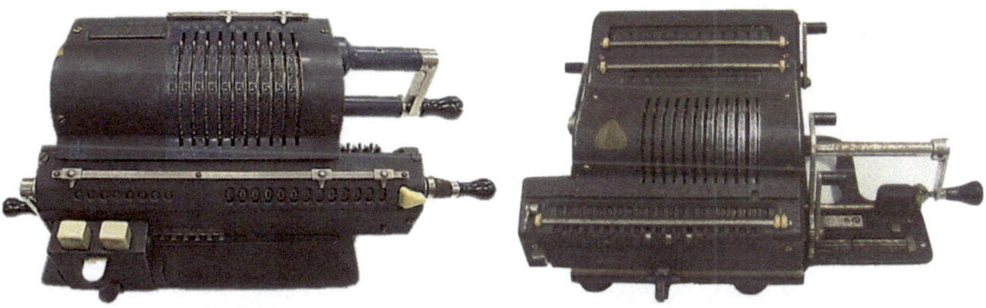

Fig. 3.2 Multiplying machines *Original Odhner* and *Brunsviga 20* (Photos: Frédérique Plantevin)

clockwise while maintaining contact with the wheel. The number 4 appears in the window corresponding to this wheel. All the digits of the two operands are entered successively, the result is read in the windows. The videos listed in the table of section "Video Accompaniment *Lightning Calculator*—LCi Files—Extract" in Appendix 4 show calculations made with this machine.

Odhner type multipliers (i.e. the original machines and all their clones built according to the same principle after the patent expired), two examples of which are shown in Fig. 3.2, perform multiplication by repeated additions. Odhner's machine is neither the first machine to do so, nor even the first to be industrially manufactured, but it is the one that has been the most produced and the most widespread, even into the 1970s.

Multipliers are first and foremost wheel adders: each order of the operands is represented by a wheel connected to a drum or to a graduated disc. But they are improved and made efficient by the addition of two parts: the driver and the carriage (Fig. 3.3). The driver is a transfer mechanism which allows to activate the wheels of all orders with a single gesture (a turn of the crank); the addition of all the digits of the two operands is therefore carried out at the same time.

The carriage allows to multiply by ten by a simple shift of the part of the machine that carries the results (called the totaliser) relative to the one that records the inscription. These two essential improvements, due to Gottfried Wilhelm Leibniz

Fig. 3.3 *Brunsviga 20* captioned (Photo: Frédérique Plantevin)

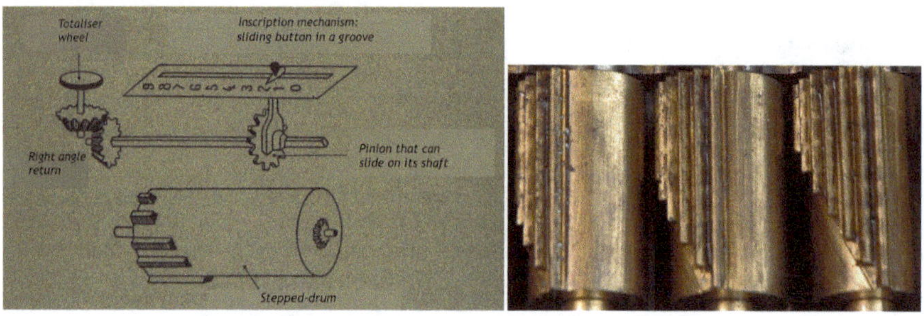

Fig. 3.4 Operating diagram of Leibniz's driver and photo of three stepped-drums (Images extracted from the panels of the exhibition "Multiply!")

as early as the end of the seventeenth century, were published more than twenty years after their conception in (Leibniz, 1710). Leibniz's driver uses a cylinder carrying grooves of unequal length called stepped-drum (Fig. 3.4). This cylinder slides on its axis depending on the choice of the inscriber. The rotation of a toothed wheel, in connection with the totaliser, meshed on this cylinder depends therefore directly on its position on it and therefore on the number entered in the inscriber. Two other types of driver will be invented afterwards, including Odhner's, known as the variable number of teeth or pinwheel. In the Odhner machines, the inscription mechanism, the so-called inscription finger, drives a cam that brings out a variable number of teeth on a wheel. The rotation of this wheel operated by the operator only drives the mechanism for the number of teeth selected by the position of the finger. You can see the teeth of the ten Odhner drivers—one per order—in the cutaway of Fig. 3.5.

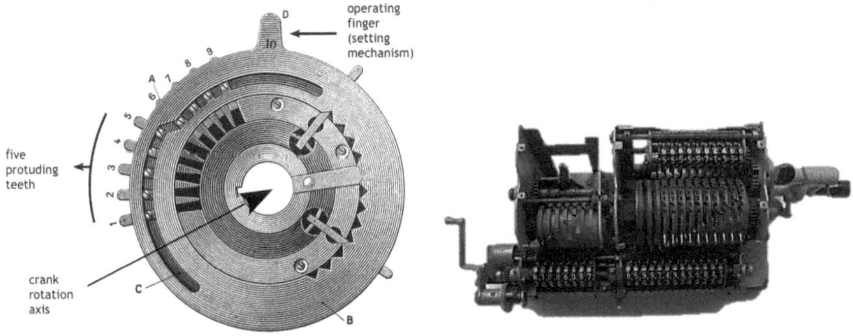

Fig. 3.5 Explanatory diagram of an Odhner driver—a cut model of an Odhner or pinwheel multiplier (images extracted from the panels of the exhibition "Multiply!")

Thus, to perform a multiplication with these machines, you must enter the digits of the first operand and turn the crank a number of times equal to the units digit of the multiplier, then shift the carriage one notch to the right and turn the crank a number of times equal to the tens digit of the multiplier, and so on until all the digits are exhausted. The result is read directly in the totaliser windows. The videos listed in the table of section "Accompanying the Videos *Brunsviga*—Bi Files—Extract" in Appendix 4 show calculations carried out with a *Brunsviga 20*.

Scientific and Pedagogical Aspects

Our sequence takes place in two stages.

The first part is dedicated to adders. One of its objectives is the understanding of how these machines work from a technological point of view, particularly in preparation for work on multipliers. It also aims at mathematical objectives. Indeed, it allows us to revisit numeration thanks to the representation of whole numbers through graduated cogwheels that these machines propose (a representation also used by multipliers). To do this, we think it useful to analyse the operation of one of these machines (whose internal mechanisms are not visible) and then to construct a prototype in order to allow pupils to truly appropriate the instrument. We can thus make the link between this representation and the operations of addition and then subtraction, which are therefore performed by the rotation of the wheels in one direction or the other, provided that the gearing allows for the automatic transfer of the carry between two orders. The concepts at play are not new for 9-year-old pupils; this revisiting of their knowledge of numeration allows, on the one hand, to understand the principle of the mechanisation of addition (a technological issue) which is based, as we have just said, on mathematical concepts, and on the other hand, to work on the representation and addition of relatively large numbers (by "adding" the necessary number of wheels) and even decimal numbers (by "adding" a physical marker representing the decimal point to the right of the wheel that one wishes to be the units'). Finally, we can work on the line-by-line writing of calculations carried out; starting from the highlighting of the properties of associativity and commutativity of addition, it leads to work on operational priorities and bracketing, the study of which is at the heart of mathematics teaching during these 9–11 years and prepares for the following years (with the properties of operations for example).

This first activity around the mechanisation of addition allows for a somewhat deeper understanding of the automation of multiplication and that of Euclidean division carried out by multipliers, which are the subject of the second part of the sequence. We can thus go beyond the simple use of a calculation instrument and access the mathematics that explain its operation. For example, the implementation of multiplication by repeated additions implemented by these machines relies on the associativity of addition and the distributivity of multiplication over addition as much as on the base 10 writing of numbers since, in order to calculate $65 \times 47 = 65 \times (4 \times 10 + 7)$, the machine actually performs

$$650 + 650 + 650 + 650 + 65 + 65 + 65 + 65 + 65 + 65 + 65 = 650 \times 4 + 65 \times 7$$
$$= 65 \times 10 \times 4 + 65 \times 7 = 65 \times (10 \times 4 + 7) = 65 \times 47.$$

Understanding the instrument from a mathematical point of view means being able to write this, which makes it an interesting tool for the chapter "Numbers and Calculations" of studies from 9 years old and even beyond 11 years old.

As for division, let's follow the same reasoning as for multiplication and write in line the calculation performed by the machine when we execute repeated subtractions of 15 from 47 for example. We get,

$$47 - 15 - 15 - 15 = 2, \text{ that is to say } 47 - 15 \times 3 = 2.$$

2 is the result of the operation, 3 is the number displayed by the lap counter; the machine rings when the operator tries to subtract 15 from a smaller number: it therefore allows to materialise quotient (3 in our example) and remainder (2 in our example) of the Euclidean division and automates the stop of the division, that is to say controls the positivity of the remainder. The work is enriched when the dividend is significantly larger than the divisor by the use of the carriage shift and naturally leads to decimal division by an integer with a number of decimals prescribed at the start.

Implementation in the Classroom

The experiment was conducted in three classes of 11-year-old pupils, over two school years and in two different ways. In one case, we had vintage calculating machines, allowing for manipulation for analysis and the execution of calculations; in the other, we used films made on purpose, without any real instrument.

It is clear to us that having working machines available to the class, ideally several, and which can be used by the pupils, is the best option in terms of motivation and interest. It allows for better pupil engagement and probably a deeper and more lasting appropriation of the activities carried out. But it is not always possible to have them. We tried to see if we could achieve the same objectives by using films showing the machines. So we prepared films, at the same time as we were designing the sessions, trying to find ways to compensate for the lack of manipulation. To our surprise, we found that the use of films, beyond being a constraint, offered new possibilities. We will show this later. In this chapter, we only present the sequence that relies on the films, but of course, we designed it based on our previous experience and we were able to see the differences between these two modalities. We first present the general plan of the sequence.

The first part on the wheel adder takes place as follows:

Understanding how the addition of numbers is "automated" by studying the adder presented above (see the statement of this session in Appendix 2), by conjecturing its internal operation and then by constructing a wheel adder that can add three-digit numbers (and whose result does not exceed three digits) with an automatic carry-over between each order. This part takes place in mathematics and technology with, if possible, moments of co-animation. In mathematics, the discovery of the adder and the beginning of the construction of the prototype represent three hours of work. The end of the construction in technology can be carried out in two hours depending on the finish and functionality that one wishes to obtain.

The second part devoted to multipliers is organised according to the following progression:

Discovering the multipliers, their functionalities by watching additions, then repeated additions, subtractions and repeated subtractions, identifying where the operands can be read; specifying how they differ from wheel adders from a mathematical/algorithmic (operations performed) and technological (way of performing them) point of view; highlighting the principle of the driver and the role of the carriage (relating it to the shift of the multiplication by a number with more than one digit in the calculation posed); if we have machines, using them to perform calculations of multiplication of fairly large numbers; proceeding in the same way for subtraction and Euclidean division; in all cases, writing in line the operations actually performed by the machine for calculating multiplication and division. This part is based on the statement proposed in Appendix 3 and represents three hours of work. It can be completed by a discussion and then a work on patents (Scherphuis, 2024) and exhibition catalogues like the one from the already mentioned exhibition (Cargou et al., 2025) for instance; this point is addressed in the "Ways to Go Further" section.

The impact of not having a real instrument is different for the two parts of the sequence. For the adder, since the pupils make a prototype themselves that they can manipulate, it is less delicate to use films to accompany the work. There is less at stake. It is even more comfortable for the teacher who can stay focused on the pupils' productions. Especially since there are explanatory videos. For the multiplier, there is no practical extension. The situation is therefore quite different. However, there are benefits to using the videos. First of all, if we have a sufficient number of videos with varied calculations, we can differentiate the activities (more or less sophisticated calculations, larger or smaller numbers), propose activities of progressive difficulty; to make the sessions more lively and allow appropriation (to compensate for the lack of manipulation), we can organise challenges of reading calculations for example. Also, the use of videos leads to a more guided work where the pupil has little initiative; on the other hand, it allows being targeted and leading to an effective and more shared institutionalisation. The sessions with manipulations are often full of effervescence, of discussions, but require a lot of attention and follow-up from teachers; the pooling of resources is often lively, but it is difficult to ensure that it is useful to all pupils in the same way, as it is difficult to know where each pupil has reached. We have in any case noticed that we asked different questions to pupils when we started working with videos. For example, it became possible to ask very specific questions, to draw pupils' attention to a particular point. Finally, the videos allow for differentiated work, as pupils can replay them as many times as necessary, in slow motion, while discussing with their partner if the session is organised in a multimedia room (with the films deposited on the computers, two pupils per station equipped with two headphones).

The films show the two machines presented at the beginning of the chapter. For each of the two machines, a brief description of the content of each film as well as its use in our sequence (objective and modality) are presented in an accompanying table (Appendix 4). The films cited in the activities are made available to teachers according to the modalities described in Appendix 5.

Discovery Session of the Adding Machine

This session is guided by a statement (Appendix 2), it is built from the films. We now follow the activity.

Highlighting the Result Windows: The Totaliser

The films allow in a first step to identify different parts of the machine and their role, the questions also allow to lift certain ambiguities or to draw the pupils' attention to some specific points that we have noticed pose problems. For example, the display dial is not a dial of inscription: it only shows the first number entered, then it shows the sum which is made simultaneously with the inscription of the next number. The fact that, at the start, the digits of the number written with the stylus appear in the windows suggests that the dial is a dial of inscription. During the session with instruments, we had noticed that some pupils were unable to perform the addition, as they entered the digits of the operands by looking at the windows and moved the wheels to make the desired digits appear in the windows. We again noticed this

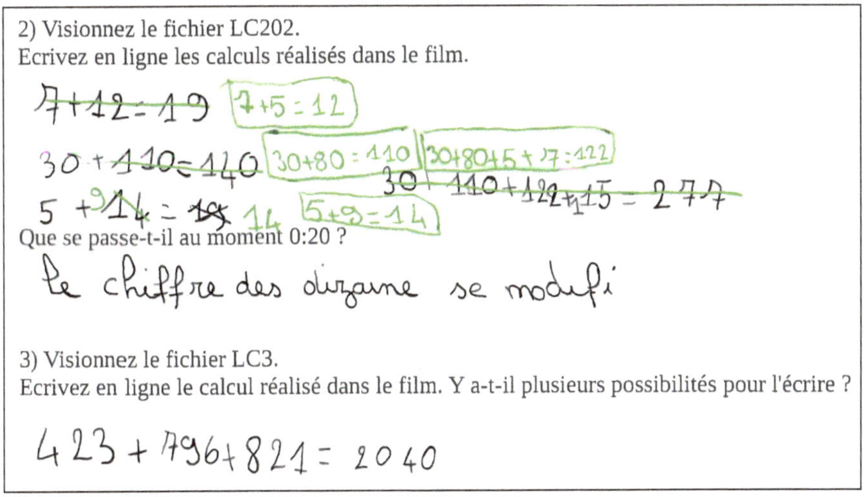
2) Visionnez le fichier LC202.
Ecrivez en ligne les calculs réalisés dans le film.

7+12=19 7+5=12

30+110=140 30+80=110 30+80+5+7=122

5+9=14 14 5+9=14

Que se passe-t-il au moment 0:20 ?

Le chiffre des dizaine se modifi

3) Visionnez le fichier LC3.
Ecrivez en ligne le calcul réalisé dans le film. Y a-t-il plusieurs possibilités pour l'écrire ?

423+796+821=2040

Fig. 3.6 Example of confusion about the number that is read in the windows: here the pupil sees them as the windows of an inscriber and not as those of a totaliser. Internal text to the image: 2) Watch the LC22 file. Write in line the calculations made in the film. What happens at 0:20? The tens digit changes. 3) Watch the LC3 file. Write in line the calculation made in the film. Are there several possibilities for writing it?

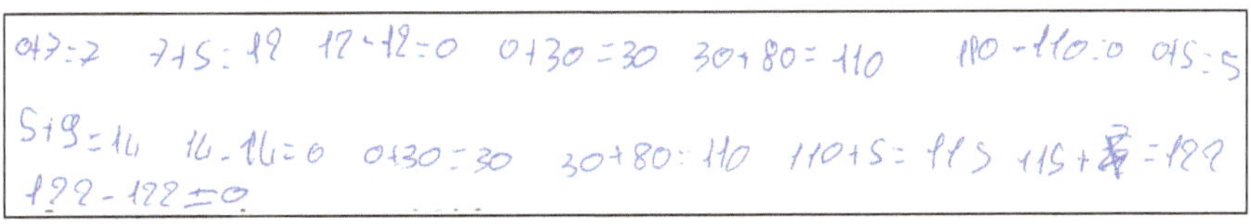
0+7=7 7+5=12 12-12=0 0+30=30 30+80=110 110-110=0 0+5=5
5+9=14 14-14=0 0+30=30 30+80=110 110+5=115 115+7=122
122-122=0

Fig. 3.7 Answers to question 2 (*cf.* the previous figure): the windows indicate the result of the addition of the number displayed in the totaliser and the number written with the stylus

difficulty in questions 2 and 3 (Fig. 3.6). In this copy, the pupil adds systematically the numbers indicated by the windows of the totaliser, without looking at the successive movements of the stylus on the wheels; in other words, he adds up the partial sums and then completes by the addition he has calculated himself.

Conversely, some pupils identify this difficulty and solve it in a very interesting way, by introducing the 0 at the start (Fig. 3.7): the windows therefore indicate the digits of the sum of the number in the totaliser (which at the start of the calculation is 0000000 = 0) and the number written with the stylus and the wheels.

It is therefore asked on very simple additions to find the operations performed: you have to look in the right place, where the stylus is placed, and not the result windows.

We wish to point out that the order in which the digits of a number are entered does not matter. For this reason, despite the fear of some of us that this may pose difficulties for pupils, we have deliberately written the digits of the operands in either ascending or descending order throughout the videos, to provoke the question.

Resetting

We want to demonstrate how the machine is reset. There are several ways to observe this: in the windows (we move the wheels so that all the windows indicate 0), or by the reverse rotation direction of the digit wheels.

This can be understood as the reverse operation of the inscription (we undo what we have just done), which is evidenced by the use of the word "*rembobine*" in Fig. 3.8, which means "rewinds"; either the machine performs a subtraction by rotating in the opposite direction (which has never been said, but which the pupils extrapolate themselves), or we only see the result of the operation which is to bring the digits in the totaliser windows back to 0. The pupils produce all possible answers. In Fig. 3.7, the pupil writes all the operations carried out according to him, including the resets which he writes as subtractions. This is an interpretation, because to perform—12 with the machine, we do not place the stylus on the 1 of the tens wheel, but on the 9, nor on the 2 of the units wheel, but on the 7 (this is what the manipulator does). The pupil extrapolates what was subtracted from 12 to bring the totaliser windows back to zero, but did not determine it by observing the machine.

In Fig. 3.8, the pupil expresses the difference between the two ways of resetting used by the manipulator in the film: for the first calculation, the stylus moves the units wheel by turning counterclockwise several times until 0 appears in the corresponding window of the totaliser. The pupil writes "*Elle a remi[s] le compteur à 0*" ("She has reset the counter to 0"). In the second calculation, the stylus is placed on the complement to 9—indicated by the red arrowhead mark (Fig. 3.10)—and brought back against the stop exactly like the inscription process, but in reverse. The pupil writes "*Elle rembobine à 0*" ("She rewinds to zero"). The vocabulary chosen by the pupil very subtly expresses the difference in method actually used. Few pupils make this remark, generally giving the same answer for the three calculations in question 2 (Fig. 3.9). One can read "She goes to 0. She resets the counters to zero. She resets the unit counters to zero".

Before proceeding with the analysis of the activity, we wish to make two remarks about resetting.

First, we see that pupils very naturally extrapolate that the adding machine performs subtraction by changing the direction of the inscription rotation; therefore,

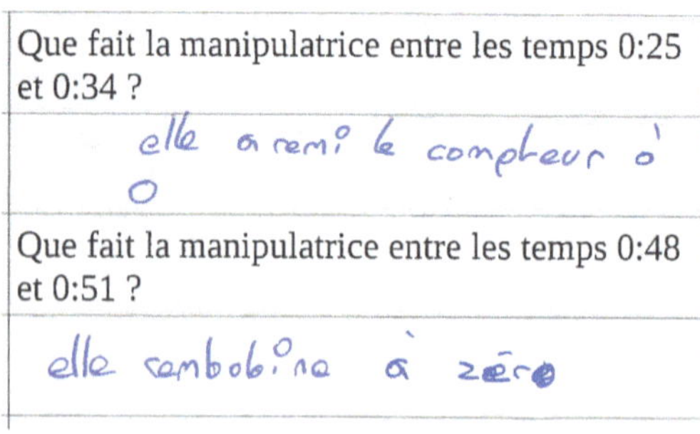

Fig. 3.8 Answers to question 2: resetting or rewinding? Text within the image: What does the manipulator do between times 0:25 and 0:34? She resets the counter to 0. What does the manipulator do between times 0:48 and 0:51? She rewinds to zero

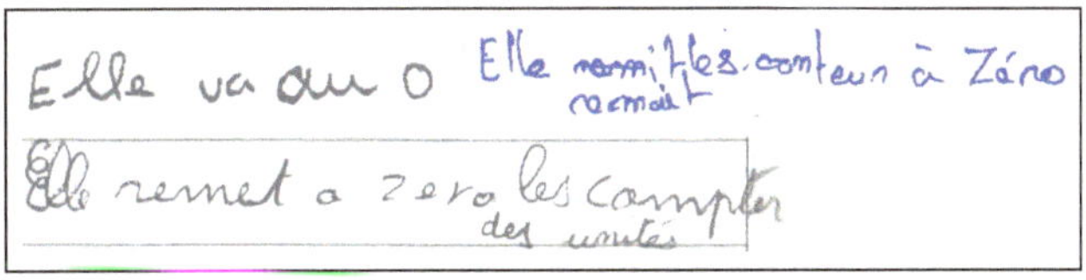

Fig. 3.9 Other examples of answers to question 2: resetting Text within the image: She goes to 0. She resets the counters to zero. She resets the unit counters to zero

Fig. 3.10 Detail of the *Lightning calculator*: reset mark (red), here underlined in white (Photo: Frédérique Plantevin)

we must answer the implicit question: can we perform subtractions in this way with this machine? It is important to do so, because pupils are convinced that it is possible when it is not (with this version of the *Lightning calculator*). A video is dedicated to this question where we show that some subtractions are possible, but not all. We can subtract from the number displayed in the totaliser a number whose each digit is less than or equal to the digit of the same order of the number displayed in the totaliser; indeed, turning the inscription wheels in the opposite direction by one notch subtracts one unit, but borrowing from the following order, however, is not possible.

Second, it may be interesting to focus on the reset mark, underlined in white in the detail of the adding machine in Fig. 3.10. This mark indicates where to place the stylus to reset to zero in one movement towards the stop (and therefore the 0) counterclockwise. The mark is actually placed in front of the 9's complement of the number indicated in the corresponding window of the totaliser. This observation, which an attentive pupil could make (although this has not happened to us yet), can lead to an interesting development (see "Ways to Go Further").

Captioned Photograph

The activity of caption of the photograph of the *Lightning calculator* comes after the work on the films. It is a synthesis activity whose essential points are the following: the identification of the windows as indicating the digits of the result that we call totaliser, the identification of the wheels as allowing the inscription of the digits of the operands, for each order of numeration, from right to left by increasing power of ten as it should be. Distinguishing these parts of the machine is essential for what follows, as is identifying the stop in order to be able to describe the inscription process clearly. The graduations and the stop allow for the writing of a user manual for this machine for non-initiates, that is, without counting or knowing anything about the mathematical operations performed (see "Ways to Go Further").

Imagining the Internal Mechanisms of the Machine

Finally, the question about the diagrams representing the inside of the machine according to the pupils comes as a conclusion of all the work of observing the machine, learning to read the results and using the machine. It also aims to prepare the work on the construction of the wheel adding machine prototype. It can even be considered the first step in the construction since, afterwards, the pupils have to combine the different wheels and discs of the material entrusted to them to achieve what the adding machine does, that is, additions with automatic carry-over. It is not only a matter of thinking about cogwheels, their gears, but also of choosing wheels that can represent numbers in our positional decimal numbering system. It produces results of a very wide variety, as shown by the pupil work in Figs. 3.11 and 3.12.

We have seen that all the pupils thought of cogwheels, which we had been able to observe during the sessions organised with the instruments in the form of free debate (Moyon & Plantevin, 2025), on the other hand for the majority of them, the number of teeth does not matter. Only one pupil chose wheels with ten teeth (Fig. 3.12). As shown in Fig. 3.11, some chose wheels with 6, 7 or 9 teeth (this last case, the most frequent, suggests that the 0 would be forgotten); others chose different numbers of teeth for each wheel for example in Fig. 3.12. If we continue on the production of pupil 1, we see that only two wheels are close enough to likely mesh. They are placed without any apparent order or concern to reproduce the studied machine, just like productions 4 and 5 of Fig. 3.11.

One can imagine that these pupils wanted to represent a gear according to them, as they have seen in technology class; which is particularly visible on diagram 4 where the directions of rotation of the different meshed wheels are indicated very precisely. Finally, some represent wheels too meshed to be able to move (production 5). Productions 2 and 3 of Fig. 3.11 propose a representation very close to the solution that will be retained for the prototype (a wheel with one tooth for the units), which is very surprising according to our experience. These pupils have probably well heard and understood the clues given by the technology teacher before the maths session of which we are studying the productions here. In the production 2, the pupil writes "When the wheel with a single tooth makes a turn, the wheel with several teeth will shift by one tooth". In the production 3, the pupil writes "As soon as we make a "click" with the units wheel, it adds 1 to the units counter" (he should have said "to the tens counter"), he indicates also "*intérieur*" ("inside") on the left, face to the first line of his diagram; it suggests that he represents both the inside and the top of the machine. It is actually quite remarkable to be able to understand

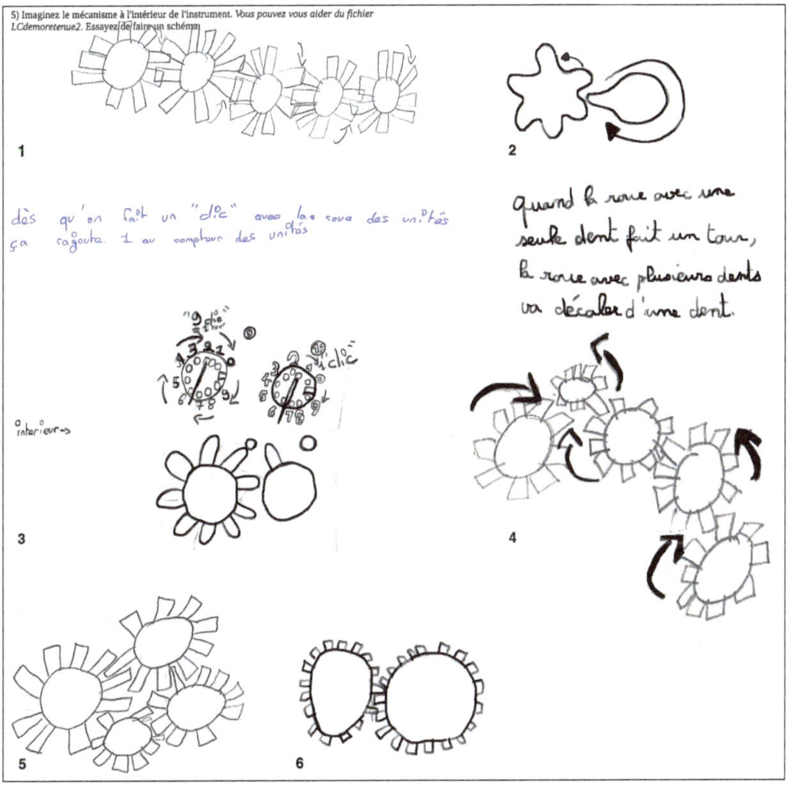

Fig. 3.11 Answers to question 5: diagrams representing the inside of the instrument

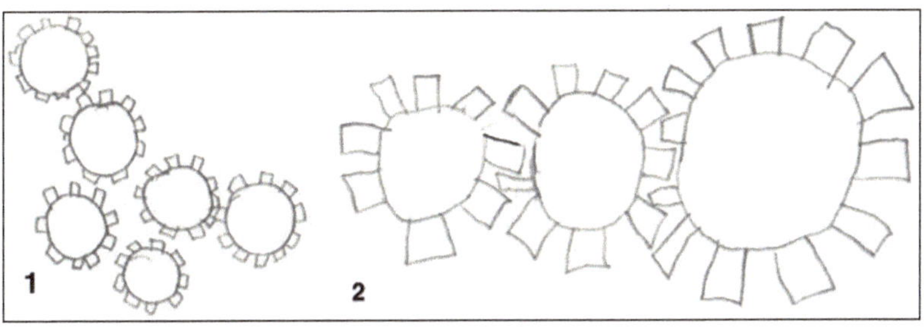

Fig. 3.12 Two diagrams representing the interior of the instrument which are impossible (with wheels equipped with a variable number of teeth, 9, 10 and 11 teeth for the first diagram, 8, 9 and 12 teeth for the second diagram)

and reproduce not only through the drawing, but also through the commentary, the technical information (arrows indicating the direction of rotation), the link between the two "strata" of the machine (production 3: totaliser windows indicated in both "strata" are with compatible scale) and even sound information. The commentary of production 2 would be nearly perfect if the wheel had indeed ten teeth and not six.

One of the objectives of the prototype construction is to correct the errors of these representations (which are therefore of a technological and mathematical nature). In particular, we wish to bring out the reasons that make the one-tooth wheel of the type of production 2, the unit wheel of an adding machine with (two) wheels: if it is divided into ten sectors (like virtual teeth), and that it is meshed as shown with a ten-tooth wheel, then it can be the unit wheel of an adding machine with (two) wheels (that is, which allows to represent the numbers (with two digits) and to add them with an automatic carryover of the remainder), because its complete turn of ten sectors will rotate the next wheel by one tooth, making at the same time of this latter, the tens wheel. The analysis of the proposed drawings is carried out in class before starting the prototype construction activity. This can be done by collecting the drawings and projecting them to the whole class to discuss the merits and defects of the proposed representations and thus prepare for what follows.

Operations and Properties of Operations

One of the objectives assigned to this discovery activity of the adding machine *Lightning calculator* is the highlighting of the properties of commutativity and associativity of addition and the use of bracketing to group calculations. Let's see how this aspect was dealt with in the activities we have proposed. In question 3, we ask to write in line the calculation made. The reason for this idea is that in these videos, the calculations are presented without comments, in particular the operands are not stated out loud. As the entry of a digit is simultaneous with its addition to the corresponding digit of the totaliser, the calculation is made as the digits are entered. The digits do not have to be entered in a certain order, as we have already said. When calculating $10 + 45 + 4$ in the LC21 video (question 1), nothing prevents writing $10 + 40 + 5 + 4$, this is indeed what the pupils generally do. In question 3, we therefore address the line writing of the calculation made in the LC3 video, calculation which is $423 + 373 + 25$ (all the numbers were entered in ascending order). As we feared that not stating the operands before the calculation would pose too many difficulties, we have prepared videos of the same calculations where the numbers are stated before being registered (table in section "Video Accompaniment *Lightning Calculator*—LCi Files—Extract" in Appendix 4) so as to be able to use them as a "boost". But we kept the statement as it was, hoping that several writings would spontaneously appear in the pupils' productions, which would have led us very naturally towards the properties of associativity and commutativity. What we obtained is represented in Fig. 3.13 (we do not talk here again about the wrong answer to this question discussed earlier).

From these answers, we can therefore start a task (for example in the form of a debate) and write the calculation of the first sum for example in all these possible forms: $423 + 373$ or $3 + 20 + 400 + 3 + 70 + 300$ or $23 + 400 + 73 + 300$ or $23+403+370\ldots$ in order to write $423+373 = (3+20+400)+(3+70+300) = (3 + 20) + (400 + 3) + (70 + 300) = \ldots$

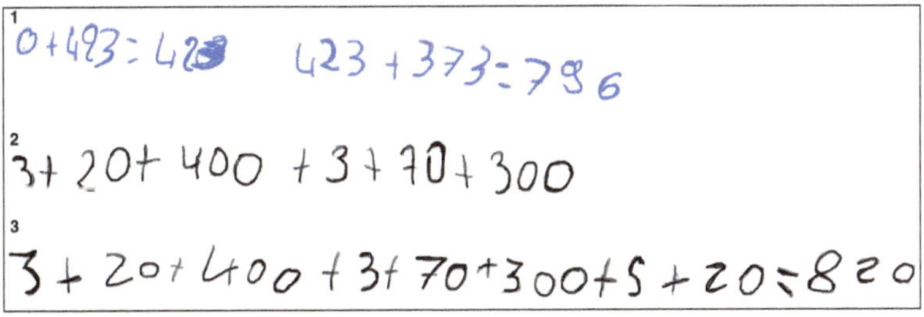

Fig. 3.13 Three in line calculations of 423 + 373 + 25 = 821 performed by the *Lightning calculator*

Reading the production of pupil 3 (Fig. 3.13), one might think that the pupil did not read the result in the adding machine's totalizer, but redid the sum himself (hence the error). Responses 2 and 3 are the most common. Pupil 1 spontaneously grouped all the digits written into a single number as long as the digits have different orders: he starts a new number as soon as a new digit arrives in the units wheel. This grouping is not obvious for pupils (pupil 1 is the only one to have done this).

Prototype Construction Sessions (Between Mathematics and Technology) In this prototype construction session, the pupil confronts a new concrete problem, his knowledge about whole numbers, their representation in base ten and about elementary operations; he must materialise these mathematical concepts in the form of a technical object that works. Mathematics and technology are therefore closely linked. That's why the session takes place in mathematics (1h30) then in technology (1h30), if possible co-animated by the teachers of both disciplines.

The activity allows to come back on the difficulties and the erroneous representations noticed in the discovery of the adding machine. The construction work is done with *ad hoc* material made available to them, in groups of three or four. The idea of this prototype is inspired by the calculating machine proposed in the old magazine *Bibliothèque de Travail* [Work Library] (Pellissier, 1965). The supplied material consists of a kit of wheels, support, nails, labels and cover. The nature, number and shape of the wheels given obviously strongly influence the pupils' design work. We will come back to this point later. Initially, they are asked to make a functional two-wheel adding machine, then a three-wheel one, and finally to take care of the finishing with the graduation and the cover.

An analysis of some of the pupils' productions is already presented in (Moyon & Plantevin, 2025). A more in-depth analysis of the activity as a whole, particularly based on the work in a previous class in the curriculum (10-year-old pupils) was carried out with a pre-made kit that had been developed for this purpose (and for teacher training). It has been published in French language in (Plantevin, 2024).

The pupils are extremely active during these sessions. Working in groups of three or four is sometimes difficult as they all want to handle the material. It is true that being able to try different possibilities is a great help (it is even impossible to do otherwise) and therefore we should try to give a kit to each pupil or at most to a pair. Especially since when working in three or four, the pupils inevitably face each other, which is problematic when it comes to placing the wheels of different orders one next to the other, a group inevitably seeing the orders reversed. One approach is to see what happens if the materials are not provided initially and it is up to the pupils to determine what they need from the available classroom materials (for example, getting the pupils to ask for ten-toothed wheels would be interesting). Many classrooms are already equipped with gear wheels that could facilitate this work (especially for primary school pupils, up to 10 years old). In addition to offering a less guided activity, this avoids the question of obtaining enough kits. We will not elaborate further on this part of our sequence.

Sessions on the Multiplier

The sessions concerning the multiplier also have several objectives. One objective, which can be described as technological, is to understand the essential elements of the operation of multipliers as an extension of wheel adders, by observing their differences in operation. This allows us to grasp the technological evolution between wheel adders and multipliers and understand Leibniz's ideas on this subject. This objective is not only technological; the way in which it can be exploited is presented below.

Multiplication by 10 is achieved by a physical shift of a part of the machine and repeated successive additions are made by as many turns of the crank as the sum of the digits of the multiplier. A "simple" adder, like the Pascaline, would compute $54 \times 47 = 65 + \ldots + 65$ $(47\ times)$ that is to say forty-seven identical manipulations each consisting of entering two digits or ninety-four gestures, while with the multiplier, it is enough to make eleven turns of the crank, plus two manipulations to enter the two digits of 65, that is thirteen manipulations. This small calculation clearly illustrates the evolution of calculating machines between adder and multiplier, and allows us to encounter, in a simple but relevant case, the notion of complexity, so important for characterising computer calculation algorithms.

Let's focus on the two purely mathematical objectives: writing down the operations performed in line and thus working on the properties of operations (commutativity and associativity of addition and multiplication, the distributivity of multiplication over addition), the instruments being then a pretext; carrying out calculations and observing how the operands are materialised, highlighting the calculation algorithms (especially with Euclidean division).

The first session takes place in a multimedia room, the pupils work in pairs in front of a computer on which the films are deposited. The statement is in

Appendix 3. We follow its progress by indicating as we go along the phases of pooling when they appeared necessary.

Unlike the adder films, those of the multiplier are generally commented; they explain the calculations made, what is shown is named. As a result, the first films serve to identify the different parts of the machine, some known, like the totaliser and the inscriber, others new like the crank. Careful observation of the films allows the legend of the photograph to be filled in, to familiarise oneself with the vocabulary in order to be able to communicate in a precise way later. The pupil's attention is drawn to the direction of rotation of the crank. The first calculations are addition calculations since this machine is actually a "super adder".

Then, we show how we multiply by repeated additions; we introduce a new dial to read: the revolution counter (of the crank) which plays the role of the multiplier in the calculation of the multiplication. In question 1 (Appendix 3), film B5 shows the two ways of multiplying by ten, with ten turns of the crank (in the direction of a clock's hands) or with the shift of the carriage (which carries the totaliser) relative to the inscriber: if the tens of the totaliser face the units of the inscriber, the unit figure of the inscribed number is added to that of the tens of the totaliser: the inscribed number has been multiplied by ten before being added to the totaliser.

To visualise things, the photos in Fig. 3.14 show details of the inscription fingers/totaliser assembly after one turn of the crank when the carriage is moved by 1, 2 and 3 notches. In Fig. 3.15, where an actual calculation is shown step by step, the oberver's attention is drawn to the revolution counter which indicates ten while only one turn of the crank has been made: thus the revolution counter no longer counts the turns of the crank. It is coupled to the carriage so that one turn of the crank made when the carriage is shifted one notch to the right counts for

Fig. 3.14 Carriage shift: the units of the inscriber are opposite the units, tens, hundreds, of the totaliser; with each turn of the crank we add 7 units, tens, hundreds to the number of the totaliser, without touching the inscription fingers (Photos: Frédérique Plantevin)

Fig. 3.15 The nine steps of the calculation of 54 × 45: 54 is inscribed, each turn of the crank adds 540 or 54 to the totaliser number depending on the position of the carriage, the "lap counter" increments by 10 or 1 depending on the position of the carriage (Photos: Frédérique Plantevin)

ten; and for a hundred when the carriage is shifted two notches to the right, and so on. We will refer to this dial as the "revolution counter" (with quotation marks since it is no longer a simple revolution counter). The first multiplication calculation involving the carriage is 54×45. It is asked to describe and then write the calculation performed, a directive supplemented orally by "write out the actual calculation performed", i.e. the calculation algorithm of the multiplication. This question is discussed collectively to ensure that all pupils have understood what the machine does and prepared question 3 on multiplications. This initially involves a precise description of the operator's movements. The answers of two pupils are visible in Fig. 3.16.

The transcription and translation of the first production (Fig. 3.16, top) gives:

> We want to find the result of 54 repeated 10 times. What is the first method? *We put 54 in the inscription fingers and then turn the crank ten times in the direction of a clock's hands.* What is the second method? *We put 54 in the inscription fingers, move the totaliser by 1 notch and then turn the crank once.* What is the last calculation performed? *54 × 5.*

The transcription and translation of the second production (Fig. 3.16, bottom) gives:

> We want to find the result of 54 repeated 10 times. What is the first method? *We enter 54 (1 turn of the crank). We then turn the crank 9 more times.* What is the second method? *We enter 54. We shift the carriage one notch to the right. We turn the crank. We move to the tens* 54 × 10 = 540. What is the last calculation performed? *We do 54 × 45/54 × 5 = 270/540/(270 + 540) + 540) + 540 = 2,430.*

To write out the calculation actually performed by the machine (video B5 and Fig. 3.15), we can use parentheses to show the grouping of calculations.

Once the calculation

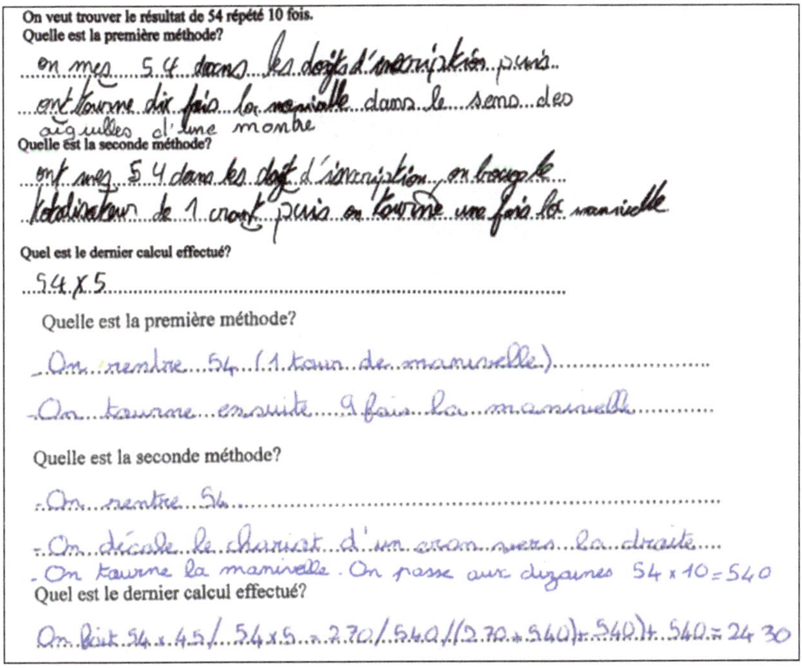
On veut trouver le résultat de 54 répété 10 fois.

Quelle est la première méthode?

on mes 5 4 dans les doigts d'inscription puis ont tourne dix fois la manivelle dans le sens des aiguilles d'une montre

Quelle est la seconde méthode?

ont mes 5 4 dans les doigts d'inscription, on bouge le totalisateur de 1 cran puis on tourne une fois la manivelle

Quel est le dernier calcul effectué?

54 X 5

Quelle est la première méthode?

On rentre 54 (1 tour de manivelle)

On tourne ensuite 9 fois la manivelle

Quelle est la seconde méthode?

On rentre 54

On décale le chariot d'un cran vers la droite

On tourne la manivelle. On passe aux dizaines 54 x 10 = 540

Quel est le dernier calcul effectué?

On fait 54 x 45 / 54 x 5 = 270 / 540 / (270 + 540) + 540) + 540 = 2430

Fig. 3.16 Pupil productions on ways to multiply by 10 with a multiplier and their implementation in the calculation of 54×45, without time for collective discussion (see below for transcription and translation)

$$54 + 54 + 54 + 54 + 54 = 270$$

is done, the multiplier performs:

$$(((270 + 540) + 540) + 540) + 540$$
$$= ((810 + 540) + 540) + 540 = (1350 + 540) + 540 = 1890 + 540$$

By using the properties of associativity and commutativity of addition, commutativity of multiplication, and finally distributivity of multiplication over addition, we find 54×45 by going through the decomposition underlying the posed calculation: $54 \times 5 + 54 \times 40$. We have:

$$(((270 + 540) + 540) + 540) + 540 = ((270 + (540 + 540)) + 540) + 540$$
$$= (270 + (540 + 540 + 540)) + 540$$
$$= 270 + (540 + 540 + 540 + 540) \text{ (associativity of addition)}$$
$$= 270 + 4 \times 540 \text{ (definition of the multiplication or distributivity)}$$

$= 270 + 54 \times 10 \times 4$ (commutativity of multiplication and calculation)

$= 270 + 54 \times 40$ (associativity of multiplication)

$= 54 + 54 + 54 + 54 + 54 + 54 \times 40$ (using the first calculation)

$= 54 \times 5 + 54 \times 40$ (definition of the multiplication or distributivity)

$= 54 \times (5 + 40)$ (distributivity of multiplication over addition)

$= 54 \times 45$

It is noticeable that in the use of the multiplier, the two operands are very differentiated: the multiplicand is registered—just like with the adder—, whereas the multiplier is constructed by the rotation of the crank on the one hand and by the movement of the carriage on the other hand. The pupils do not seem to be troubled by this, but it would probably be necessary to pay attention to this issue with younger pupils and ensure that it does not induce an intuition against the commutativity of multiplication.

Subtraction poses no difficulty for the pupils. After having described the calculation carried out in the film, we ask to describe a similar calculation done mentally with the machine, to ensure that things are understood. A pupil's response is shown in Fig. 3.17 whose translation gives:

> *We display* 2,881 *on the registration fingers we turn the crank once clockwise then we place* 1,429 *on the registration fingers and we turn the crank counterclockwise.*

After this question, there is only one feature of the multiplier left to discover, it is the bell. It rings when the operator tries to subtract a number larger than the one indicated in the totaliser. This device is very practical for the implementation of the division calculation since it serves as a test of positivity of the remainder in the Euclidean division. We will see this below. The following questions implement the different features of the machine in real multiplication and division calculations.

The last question is about Euclidean division. First a naive division by repeated subtractions is presented, that is, without using the carriage. An order of magnitude of the number of crank turns is asked for. We have chosen a dividend (3725) large compared to the divisor (14), the quotient is 266, which represents a lot of crank turns to execute, as shown in film B8. This motivates the learning of the expert

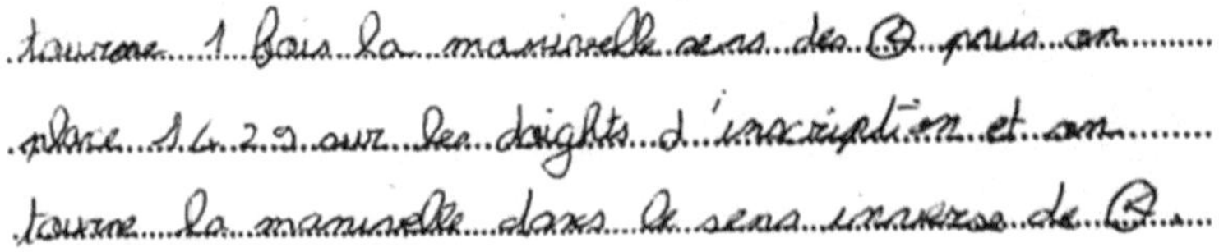
On affiche 2881 sur les doights d'inscription on
tourne 1 fois la manivelle sens des [illegible] puis on
place 1429 sur les doights d'inscription et on
tourne la manivelle dans le sens inverse de [illegible]

Fig. 3.17 Mental calculation of the subtraction 2881 − 1429

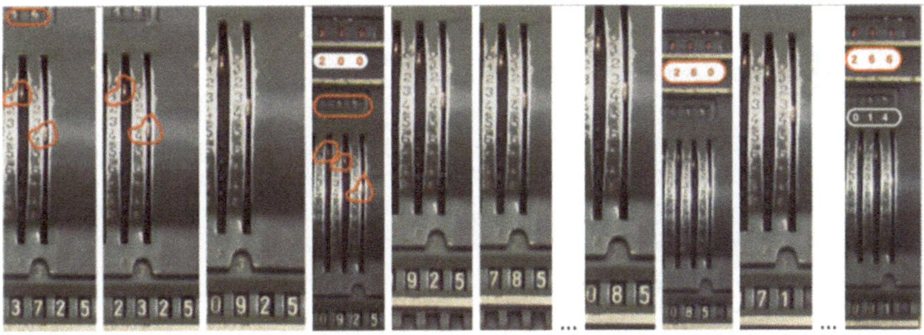

Fig. 3.18 Ten of the fourteen steps of the calculation of the Euclidean division of 3725 by 14: 3725 is registered, 14 is registered; each turn of the crank subtracts 1400, 140 or 14 depending on whether the carriage is shifted by two notches, one notch or is in its initial position. The “lap counter” increments by 100, 10 or 1 depending on the position of the carriage (Photos: Frédérique Plantevin)

method implemented in Fig. 3.18. It consists of subtracting 1400 from the dividend (instead of 14), by shifting the carriage two notches to the right, then subtracting 140 as many times as possible then 14. The “lap counter” therefore gives the quotient and the totaliser indicates the remainder.

An example of the work of two pupils is shown in Fig. 3.19. Although they forgot the remainder in the line writing of the last question, they understood the activity well after their independent work. Few pupils achieve this result without a step of collective discussion; probably it would be useful to carry out the calculation in parallel and also, in another order of idea, to start with smaller numbers and pairs of dividend and divisor of the same order of magnitude.

We can also write the calculations performed online, as we have done for multiplication. Reviewing the calculation in parallel, as the pupils in Fig. 3.19 began to do, seems a natural idea at this stage. The multiplier offers a materialisation of the division calculation, of its operands and result, which can help pupils better understand and remember its algorithm for its calculation.

Ways to Go Further

The work proposed with the multipliers can be extended in various ways in mathematics classes, but also in technology and possibly history class. Let’s start with mathematics. Once the pupils are familiar with the operating of these machines, we can deepen the division of integers by considering divisions with decimal quotients. We must decide the number of decimals we want to have at the beginning of the calculation to shift the carriage by as many notches when entering the dividend. Demonstration films are available for eventual use in class (B10 and B11,

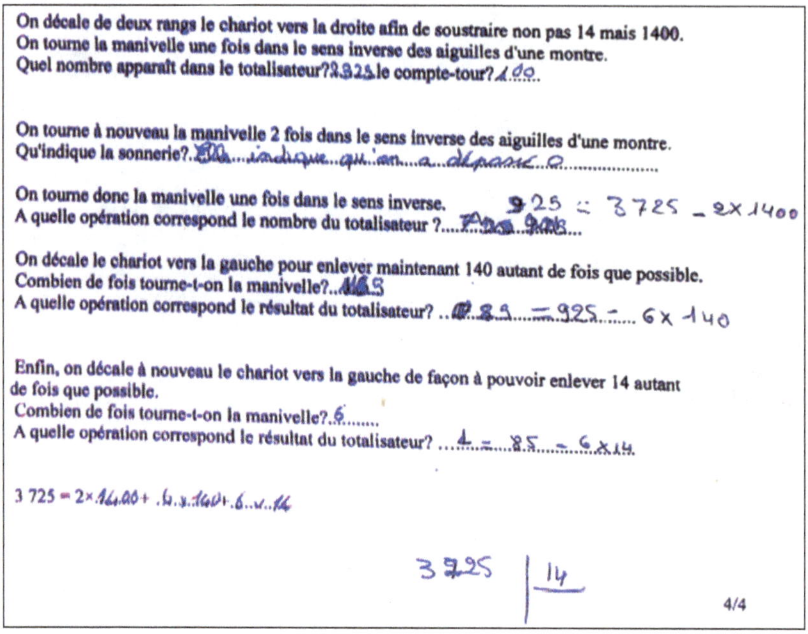
On décale de deux rangs le chariot vers la droite afin de soustraire non pas 14 mais 1400.
On tourne la manivelle une fois dans le sens inverse des aiguilles d'une montre.
Quel nombre apparaît dans le totalisateur?2325 le compte-tour?100

On tourne à nouveau la manivelle 2 fois dans le sens inverse des aiguilles d'une montre.
Qu'indique la sonnerie?... indique qu'on a dépassé 0

On tourne donc la manivelle une fois dans le sens inverse. 925 = 3725 - 2x1400
A quelle opération correspond le nombre du totalisateur ?...

On décale le chariot vers la gauche pour enlever maintenant 140 autant de fois que possible.
Combien de fois tourne-t-on la manivelle?...
A quelle opération correspond le résultat du totalisateur? ...89 = 925 - 6 x 140

Enfin, on décale à nouveau le chariot vers la gauche de façon à pouvoir enlever 14 autant de fois que possible.
Combien de fois tourne-t-on la manivelle?6
A quelle opération correspond le résultat du totalisateur? ...1 = 85 - 6x14

3 725 = 2×1400 + 6x140 + 6x14

3725 | 14

4/4

Fig. 3.19 An answer to the questions on division (the statement in English is in Appendix 3)

cf. table in section "Accompanying the Videos *Brunsviga*—Bi Files—Extract" in Appendix 4).

Zeroing the adder can also lead to a work on the complement to nine and lead to an interest in the way in which the *Pascaline* performed subtractions. We can also be interested in adders of the type of the *Lightning calculator* which allow subtraction and therefore have a double graduation. It could then be proposed to graduate the prototype to allow it to make subtractions and to "compare in thought" thus the two instruments.

For both machines, the sequence can lead to the creation of a user manual intended for people who have no knowledge in mathematics and do not know how to calculate. This is a way of getting pupils to state a procedure, very close to the concept of algorithm.

These activities lead to numerous developments in technology or between technology and mathematics. As mentioned above, the sequence is designed so that pupils understand the key elements of the multiplier and what differentiates it from an adder. This can be done in the form of a debate to highlight the differences between adder and multiplier; we successfully tested this in class (in the modality where we had machines available): the pupils saw that the multiplier calculates all the digits of the totaliser at the same time, which leads to the concept of driver. Leibniz's extraordinary idea is better understood. More details on the evolution of multiplying machines can be found in Marguin (1994), Plantevin (2018) and

Bruderer (2020). The analysis and description of technical objects, or even the design of a technical object to meet a need, are at the heart of science and technology teaching for pupils between 9 and 11 years old. Both are addressed by the proposed activity. The comparative study of different clones of Odhner multipliers can introduce the concept of technological evolution. It can also provide an opportunity to present to pupils the patents of the machines used, which are now available on the web. In particular, it seems natural to conclude the development of the adder prototype by a study of the patent of the *Lightning calculator* to find out how it is made.

The industrialisation of production methods is a major component of the history of these machines. It was Charles-Xavier Thomas de Colmar who first improved Leibniz's machine in terms of functionality and reliability, but also so that it can be manufactured using industrial processes. This development work took him 31 years, between his first patent (1820) and the start of industrial production of his arithmometer (1851). Studying his journey and his work on his machines is a relevant illustration of the Industrial Revolution. One can also be interested in the evolution of production methods in some of the factories manufacturing clones of Odhner machines during the twentieth century (for example Brunsviga, for which it is relatively easy to obtain various resources). For the pupils with whom the activity was tested, the sequence contributed to the establishment of historical markers, which is a permanent objective of history teaching.

Supporting the teaching of mathematics, and in particular of numeration and arithmetic operations, through the study of old or obsolete calculating instruments as we do, is not unique. The compilation work of Volkov and Freiman (2019) testifies to the diversity of possible artefacts and techniques and therefore of varied classroom activities that result from them. More specifically, the works of Maschietto (2013) and Poisard (2006) are interested in the didactic aspects of the use of an adding machine artefact made for the class (pedagogical material *ad hoc* made upstream) and called Pascaline, for the former, and abacuses for the latter (there are now digital online versions of them). Their questions are close to ours and therefore partly illuminate our work. However, in the activity we propose, we work on historical instruments (and not artefacts constructed for the class) and above all, we ask pupils to design their own prototypes of machines, within an activity partly mathematical. We therefore seek to lead pupils to understand these machines as technical objects embodying mathematics that they make explicit (and not only to use them for performing calculations). We have been able to observe that the activity allows some pupils to show a kind of intuition or practical intelligence by manipulating and creating technical objects themselves. It seems to us that this talent is a possible springboard for learning the associated mathematics and that it should be encouraged by well-chosen activities.

Finally, this dual perspective, between mathematics and technology, both in the approach used in the activity and in its meaning, is very specific. It is also original (as far as we know).

Acknowledgments This chapter was written by Frédérique Plantevin based on the work of the IREM research group in Brest (France) "Instruments in history and in the classroom". Véronique Fustec, Yohan Lefèbvre, Yann Le Thénaff from Saint-Pol-Roux college in Brest and Marc Le Pors from the Brest city district participated in this group between February 2017 and June 2018. Murielle Geslin from the Champ-de-Foire school in Plougastel-Daoulas participated in the group between February and June 2017. Dominique Bernard from the Jacques-Prévert school in Brest joined the group in the autumn of 2017.

Appendix 1: Concepts and Skills Addressed in This Chapter

Covered Concepts

- Using and representing large whole numbers..., decimal numbers
  - Understanding the rules of operation of numeral systems in the field of whole numbers to represent them with cogged and meshed wheels; application to large numbers according to the capacity of machines between 10 and 12 digits; extension to decimal numbers by materialising the decimal point.
- Calculating with whole numbers and decimal numbers
  - Properties of operations (commutativity, associativity, distributivity of multiplication over addition).
  - Line calculations: using parentheses (to describe the operations performed by the multiplier).
  - Implementing a calculation algorithm for multiplication and division in relation to mechanised calculation.
  - Working on the meaning of operations, particularly division (role and representation of dividend, divisor, quotient and remainder).
- Materials and technical objects
  - Identifying the evolution of needs and technical objects in their context: from the mechanical calculation of addition to that of multiplication.
  - Describing the operation of technical objects, here adder and multiplier, their functions and their components.
  - Designing and producing a technical object (wheel adder) as a team to translate a technological solution that meets a need (automating the carry in addition).

Developed Skills

- **Search and research skills**: Manipulate, experiment from new supports.
- **Computational skills**: Discover and use a new operational technique (mechanised calculations) in connection with traditional calculation techniques (in line, posed).
- **Mathematical communication skills**: Argue in exchanges (around the construction of a prototype in pairs). Report observations, hypotheses, conclusions using precise vocabulary.
- **Mathematical representation ability**: Use machines to represent numbers: represent orders by wheels, digits by teeth, carry by means of a gear..., especially operands in different operations (entered operands, constructed operands).

Appendix 2: Statement of the session Discovery of the Adding Machine *Lightning Calculator*

Warning: The file names in the appendix, as well as throughout the chapter and the pupil productions refer to the original ones in French language. The English version of the film named "*filename*" is the film named "*filename_eng*".

1. Watch the LC21 file. Fill in the following table:

Write down the calculation performed between times 0:08 and 0:18	What does the operator do between times 0:25 and 0:34?
Write down the calculation performed between times 0:38 and 0:44	What does the operator do between times 0:48 and 0:51?
Write down the calculation performed between times 0:55 and 1:20	What does the operator do between times 1:25 and 1:31?

2. Watch the LC22 file. Write down the calculations performed in the film. What happens at time 0:20?
3. Watch the LC3 file. Write down the calculation performed in the film. Are there multiple ways to write it?
4. Name each part of the instrument. Try to choose a name that explains what this part does. *You can use the films for help.*

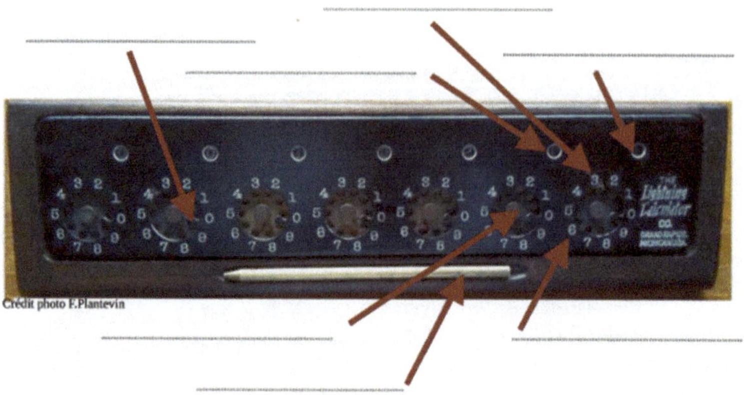

5. Imagine the mechanism inside the instrument. *You can use the LC6 file for help.* Try to draw a diagram.

Appendix 3: Statement of the session on the Multiplier Machine

Start by watching the B0 video.

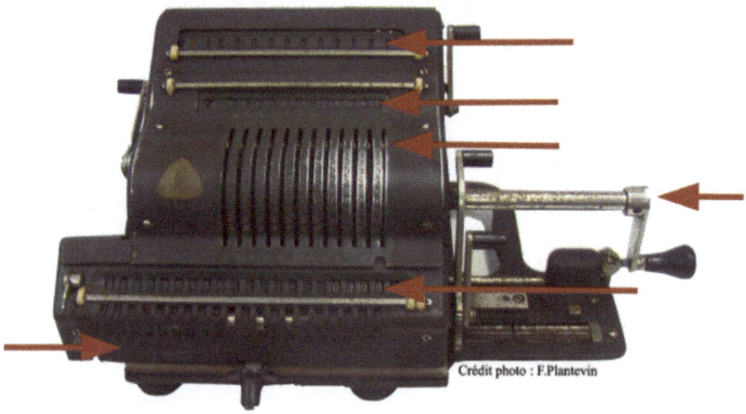

You will complete the caption of the photo above as you progress through the activities.

1. **Addition** (videos B2 and B3)
   Example: 542 + 338
   To display the number 542 in the register, we use input fingers. We put the unit finger on the number 2. We put the tens finger on the number 4. We put the hundreds finger on the number 5. We turn the crank clockwise so that 542 appears in the totaliser.

To add 338, we put the finger of $\cdots$ on the number $\cdots$, we put the finger of $\cdots$ on the number $\cdots$, we put the finger of $\cdots$ on the number $\cdots$. We turn the crank again in the direction $\cdots$ What number appears in the totaliser?
A first remark (video B4):
We add the number 333 several times. What is the name of this new dial mentioned in the video? $\cdots$ This machine therefore has at least one other function besides addition. What is it?
A second remark (video B5):
We want to find the result of 54×10. What is the first method? What is the second method? What is the last calculation made?

2. **Subtraction** (video B6)
   Example: $2821 - 774$
   We display the number 2821 in the register using the registering fingers. We turn the crank clockwise. What do we do next to subtract 774 from 2821? What number appears in the totaliser?
   It's your turn to calculate!
   Explain to us in writing how to get the result of $2881 - 1429$, detailing all the steps.
3. **Multiplication** (video B7)
   Example: 2856×563
   We display the number 2856 in the register using the registering fingers. We turn the crank clockwise.

   - We first multiply 2856 by 3 (units digit). What do we do? What number appears in the totaliser?
   - We shift the carriage one place to the right. We turn the crank $\cdots$ times clockwise. What number appears in the totaliser? What operation corresponds to the result indicated on the totaliser?
   - We shift the carriage one place to the right. We turn the crank $\cdots$ times clockwise.

   What number appears in the totaliser? What operation corresponds to the result indicated on the totaliser?
   Complete the calculations in line:
   $2856 \times 563 = 2856 \times 3 + \cdots$
4. **Division** (videos B8 and B9)
   Example: $3725 \div 14$
   We display the number 3725 in the register using the registering fingers. We turn the crank clockwise. We reset the counter. We display 14 and turn the crank counterclockwise as many times as necessary (video B8). We can notice that it's very, very long! Give an order of magnitude of the number of crank turns? $\cdots$.
   Give the quotient $\cdots$ and the remainder $\cdots$
   Example: another more reasonable method (video B9)!
   We display the number 3725 in the register using the registering fingers. We turn the crank clockwise. We reset the counter. We display the number 14 in the register using the registering fingers. We shift the carriage two places to the right

in order to subtract not 14, but 1400. Turn the crank once anticlockwise. What number appears in the totaliser? $\cdots$, in the counter? $\cdots$
Turn the crank again twice anticlockwise. What does the bell indicate?
So, turn the crank once anticlockwise. What operation does the number on the totaliser correspond to?
Finally, shift the carriage back to the left in order to remove 14 as many times as possible. How many times do we turn the crank?
What operation does the totaliser result correspond to?
Complete the calculations in line:
$3725 = 2 \times \cdots + \cdots + \cdots$

Appendix 4: Film Accompaniment Tables

Video Accompaniment *Lightning Calculator*—LCi Files—Extract

File name	Description	Use
LC0	Presentation of the instrument (date of manufacture, manufacturer, patent).	**To start in whole class.**
LC1	Presentation of two additions of three-digit numbers.	To show the instrument in action.
LC21	Calculations without commentary $5 + 9$, $7 + 2$, $10 + 40 + 5 + 4$ with reset to 0 using or not the mark (red angle) of the complement to 9.	**For the pupil's independent work.** To be placed on the pupils' workstation for pair work. Discovery of the instrument in action.
LC22	Calculations without commentary $7+5$, $30+80$, $5+9$, $30+80+5+7$.	
LC3	Calculations without commentary $423 + 373 + 25$ and reset to 0.	
LC31	Calculations without commentary from LC3, but with the numbers stated before the calculation; calculations redone a second time.	**A boost for independent work.** Variant of LC3. If the pupils cannot find the numbers written after the first.
LC5	Explanation at the beginning. Numbers are stated, with time in between to allow time to pause. 234 + 427 + 106 and reset to 0.	**Sharing** after the two-person investigation phase. The film is video-projected on the board, the teacher pauses after each number stated to ask the pupils: how to do it?
LC6	Explicit calculations: the operator says what she is doing, how the calculation is conducted (direction of rotation, stylus in front of, until the stop); several parts of the instrument are named; 11 + 22 + 18.	**Carryover**. *This file is for the teacher, but can be shown to pupils to focus on the automatic carryover in addition.*

Accompanying the Videos *Brunsviga*—Bi Files—Extract

File name	Description	Use
B0	Presentation of the instrument, date of manufacture and manufacturer mentioned, patent.	**To start in whole class.**
B2	Presentation of an addition calculation explained with the vocabulary $542 + 338 = 880$.	**For pupils to work independently**.
B3	Some additions explained: $62 + 546 = 608$ and $608 + 131 = 739$.	
B4	$333 + 333 + 333$ discovery of the counter, 3 repeated additions "Any adding machine can be a multiplier by repeated additions" but this one is very special we will see why later.	**To be placed on the pupils' workstation** (tablet, computer) for pair work. Identify the instrument, how to use it and notice what it does.
B5	Calculations without commentary $54 + 54 + 54 + 54 + 54 + 54 + 54 + 54 + 54 + 54 = 540$ (10 turns of the handle), $54 \times 10 = 540$ (one notch shift of the carriage to the right) then calculation of $54 \times 45 = 2430$.	Multiplication by 10. Discovery of the carriage without commentary 2:34.
B6	$2821 - 774 = 2047$ and $2047 - 774 = 1273$	**Presentation of subtraction**.
B7	$2856 \times 3 = 8568$; $8568 + 28{,}560 \times 6 = 179{,}928 = 2856 \times 63$; $179{,}928 + 285{,}600 \times 5 = 1{,}607{,}928$	**Complex multiplication**.
B8	Division of 3725 by 14 with 266 crank turns, role of the revolution counter dial and the totaliser.	**Presentation of division** in a naïve way, directly by repeated subtractions; 4 minutes of crank turns.
B9	Calculations without comments $3725 \div 14$.	
B10	Calculations without comments, division with two decimals $3725 \div 14$.	**Demonstration to go further**: decimal division
B11	Calculation division with two decimals - with comments $54 \div 21$.	**Demonstration to go further**: decimal division

Appendix 5: Access Modalities to Online Resources

The files freely accessible on the web of this chapter are gathered on the presentation page of the French language book on the national IREM site from the URL: https://www.univ-irem.fr/-passerelles-enseigner-les-mathematiques-par-leur-histoire-en-cycle-3- and more precisely on the page of the corresponding chapter 3.

The links to the videos are available for viewing or downloading on this same page, or directly on https://ubotv.univ-brest.fr/search/?q=plantevin and then to apply the langage filter "anglais" to get all the available videos.

References

Bruderer, H. (2020). *Milestones in analog and digital computing*. Springer. https://doi.org/10.1007/978-3-030-40974-6

Cargou, C., Cargou, M.-P., & Plantevin, F. (2025). *Multiply! Instruments and machines that multiply. 200 years of strokes of genius, 200 years of industrialisation*. IREM de Brest.

Hoyau, L.-A.-D. (1822). Description d'une machine à calculer nommée Arithmomètre, de l'invention de M. le chevalier Thomas, de Colmar. *Bulletin de la Société d'encouragement pour l'industrie nationale*, *221*, 355–365.

Leibniz, G. W. (1710). Brevis description machinæ arithmeticæ, cum figura. In *Miscellanea Berolinensia ad incrementum scientiarum* (pp. 317–319). Deutsche Akademie der Wissenschaften zu Berlin, Berlin. Translated into French by Colette-Chantal Adam. https://doi.org/10.4000/bibnum.551

Marguin, J. (1994). *Histoire des instruments et machines à calculer. Trois siècles de mécanique pensante, 1642–1942*. Hermann.

Maschietto, M. (2013). Systems of instruments for place value and arithmetical operations: An exploratory study with the Pascaline. *Education*, *3*(4), 221–230. https://doi.org/10.5923/j.edu.20130304.02

Moyon, M., & Plantevin, F. (2025). Teaching mathematics through historically-based activities: Two educational experiments. *HISTEMAT. Revista De História Da Educação Matemática*, *11*. https://www.histemat.com.br/index.php/HISTEMAT/article/view/758/576

Pellissier, M. (1965). Construis une machine à calculer (addition et soustraction). *Bibliothèque de Travail*, *612*(supplément), 189.

Plantevin, F. (2018). Les génies de la multiplication. In E. Barbin, D. Bénard, & G. Moussard, (Eds.), *Les mathématiques et le réel : expériences, instruments, investigations* (pp. 111–125). Presses universitaires de Rennes.

Plantevin, F. (2024). Roues dentées et engrenages, nombres et opérations : allers et retours. *Repères-IREM*, *136*, 5–30. https://irem.univ-grenoble-alpes.fr/revues/reperes-irem/consultation-en-ligne/

Poisard, C. (2006). The notion of carried-number, between the history of calculating instruments and arithmetic. In P. Groontenboer, R. Zevenbergen, & M. Chinnapan (Eds.), *Identities, Cultures and Learning Spaces, Proceedings of the 29th Annual Conference of the Mathematics Education Research Group of Australasia* (Vol. 2, pp. 416–423). MERGA.

Scherphuis, J. (2015–2024). The Calculator/Lightning Adding Machine & Original-Odhner. In the website: The Jaap's Mechanical Calculator Page. https://www.jaapsch.net/mechcalc/lightning.htm & https://www.jaapsch.net/mechcalc/odhner.htm

Volkov, A., & Freiman, V. (2019). *Computations and computing devices in mathematics education before the advent of electronic calculators, mathematics education in the digital era series*. Springer. https://doi.org/10.1007/978-3-319-73396-8

Frédérique Plantevin is mathematician at the University of Brest (France), member of LMBA-CNRS UMR 6205. Through her involvement in national and international networks (IREM, SIC), she has developed a line of work on historical instruments in primary, secondary and higher education classes of mathematics. She is interested in the study of mathematical instruments, in particular those based on tractional motion. She founded the Cabinet of Curiosity in the Faculty of Science where the collection of outdated scientific instruments is housed and participates to its enhancement.

Chapter 4
Number Ratios: Another Approach to Fractions

Sylviane R. Schwer

Monstrat ars numeris que virtus possit habere.
The art of number shows what virtue it can have.
Explico pernumeru que sit proportio rerum.
I explain by number what is the proportion of things.

(Tapestry Arithmetic, *circa* 1520, Cluny Museum, France)

Abstract Number relationships are at the heart of the Pythagorean thought from Greek Antiquity to the Middle Ages, in particular with Boethius, and even up to the seventeenth century. The essential difference between a number and its various representations is highlighted. For example, a fraction is just one of many representations of a rational number. The activities also provide an opportunity to work on the concept of classification (an important scientific practice) and the vocabulary associated with the cardinal and ordinal characteristics of integers.

Keywords Number's meanings · Number's representations · Number's properties · Number's relationships · Ratios Greek classification and naming

Introduction

Until the beginning of the ninth century, in Europe, the teaching of arithmetic mainly concerned the learning of elementary operations, using abacuses with tokens, for young monks and young merchants. Rediscovering Boethius' *Arithmetic* and *Music* allowed monks and scholars to rediscover the Pythagorean approach to number, more focused on the properties of numbers themselves, like even/odd, or on the

S. R. Schwer (✉)
IREM Paris Nord, Université Sorbonne Paris Nord, Villetaneuse, France
e-mail: schwer@math.univ-paris13.fr

M. Moyon et al. (eds.), *Teaching Mathematics Through Historically-Based Activities*, History of Mathematics Education,
https://doi.org/10.1007/978-3-032-16990-7_4

relationships that bind them, or *ratio*. Boethius's treatise of arithmetic is the only Neoplatonic treatise known in the Middle Ages. It is a Latin adaptation of the Greek text of Nicomachus of Gerasa's *Introduction to Arithmetic*. It played a crucial role in the mathematical education of the elites of the Middle Ages (Gibson, 1982; Høyrup, 2014; Getenet & Callingham, 2017).

In this chapter, we propose to follow the historical perspective of the concept of natural numbers and then fractions through Pythagorean thought and its transmission, which served as the basis for academic teaching of arithmetic in the Middle Ages and which is commonly found in the sixteenth century (Recorde, 1557), and even up to the seventeenth century (Jean Abraham, known as Launay, 1665) and eighteenth century in musical treatises (Fux & Miller Schulte, 2017). This approach allows, with the help of simple manipulations of tokens, to approach the concept of fraction without handicapping weak pupils in calculation.

The approach also meets an essential need for the development of scientific thought and the education of the citizen, that of studying an object or a problem from different points of view and choosing the best one for the situation at hand. Establishing correspondences between different representations of the same object, understanding what is specific to the object and what depends on the chosen representation is the main objective. Numerous didactic and pedagogical studies have been published concerning the learning of fractions, for example (Brissiaud, 1998). It is still a difficulty not overcome by many pupils aged 11 and over who often only know, as introductions to fractions, sharing activities. This chapter proposes to play with an old—and, I believe, necessary—vision of fractions to better manipulate them.

From a purely mathematical point of view, this approach allows us to work on the essential difference between a number and its various representations. A fraction is just a representation of a number qualified as rational. The Greeks were able, through this approach, to give a name to the rational number, which the French or English language—as most languages—does not do, apart from multiplicity ratios. We can say that the pairs (20;5) and (40;10) are quadruple ratios, but it takes a sentence to express the equivalence of the ratios between 21 and 5 on the one hand, and 42 and 10 on the other. We can say that 21 is to 5 as 42 is to 10. The Greeks will give them a common ratio name, which indicates how to position it exactly on the number line.

In this chapter, we propose a sequence of activities, also allowing to work on scientific practices of classification, on vocabulary related to the cardinal and ordinal characters of integers, and on the coding and decoding of expressions qualifying number ratios, which directly meet the institutional and scientific requirements of knowing how to position a fraction on the number line. This sequence extends a project carried out at the Orméteau school in Aulnay-sous-Bois (a town in the northern suburbs of Paris, France) which we will report on in the implementation paragraph.

Historical Context

In Greco-Latin antiquity, and up to modern times, a mathematician is both a philosopher and a physicist or engineer. The main questioning of the philosophers of Antiquity concerns the rational explanation of the universe, including its birth. Several theories are defended. The two main and often opposed sources of Greek thought are represented on the one hand by the eternalist conception of Parmenides of Elea (*circa* 504–450 BCE), and on the other hand the river conception of Heraclitus of Ephesus (*circa* 550–480 BCE).

Heraclitus is a pre-Socratic contemporary of Pythagoras (580–495 BCE). His reflections focus on the nature of the universe, the sensible world, what is called physics. For him, "nothing is, everything becomes" through a perpetual and alternating movement of opposites: night/day, hot/cold..., which is illustrated by the maxim "You cannot bathe twice in the same river". Parmenides, considered the father of *metaphysics*—that which is beyond the sensible world—is interested in what is the raw material of the universe, the world of Truth, that which is eternal, ungenerated, and perfectly immobile. The space in which we live is a mere reflection of this world of Unity. Any perception of the sensible world can only be tainted with error. "Our senses deceive us," as Montaigne (1533–1592) liked to remind us. Beyond this is a world of Reason, of Truth, of Permanence. The story of Theseus's ship is a pleasant illustration of this question of identity through time. The ship on which Theseus had embarked with other Athenians to fight the Minotaur returns victorious to Athens. The Athenians preserved this ship for a very long time, changing the parts—identically—one by one as they deteriorated. A day came when not a single original piece remained. The question that philosophers still debate is whether the ship is still Theseus's ship or if it is another vessel? The answer depends on what makes the identity of the ship: its constituents or its *raison d'être*.

It is from this substrate that Pythagoras and his school emerge, for whom the number—natural integer, the one that translates the plurality of things and corresponds to the question how many?—"is the essence of all things". The number then becomes both representation and explanation of the world. Aristotle will define Time as "the number of movement according to the prior and the posterior". Alongside numerology, the study of numbers in themselves occupies an important place in the Pythagorean school. It will be a matter of studying numbers as beings in themselves: according to their characters or properties and according to the relationships they maintain with each other. These properties and relationships are elaborated from a representation of the number as a collection of unit-tokens, and arrangements of these units in geometric forms: segments, triangles, squares, cubes...

Pythagoras had his political heyday and then his fall. His school functioned like an elitist sect and on the basis of secrecy. A rejected individual set fire to the school building on a meeting night, which led to the death of most of the disciples. It is through the criticisms of Plato (*circa* 428–348 BCE) and Aristotle (384–322 BCE) that Pythagorean thought survived and was rekindled, tinged with new philosophies. The Greek and neo-Pythagorean Nicomachus of Gerasa (*circa* 60–120 CE) took up the torch in his *Introduction to arithmetic*, then thanks to the Latin translation (Boethius, 1983) made by Boethius (*circa* 480–524), the Middle Ages and the European Renaissance will inherit it (Gibson, 1982; Høyrup, 2014; Caiazzo, 2021). In the Middle Ages, it was the Church that determines the subject matter of university knowledge. It took up the division of letters and sciences from Antiquity into seven liberal arts. Grammar, dialectic, rhetoric form the *trivium* and constituted the first degree for understanding the Scriptures. The second degree, called *quadrivium*, consisted, in the Pythagorean order transmitted by Boethius, of arithmetic, music, geometry, and astronomy. These four disciplines were the means of glimpsing how God had organised the world. These liberal arts formed the basis of university studies. Each of these arts was the subject of iconography and in particular of a sixteenth-century tapestry, *Arithmetic*, which is located in the Cluny Museum (National Museum of the Middle Ages, Paris, France). The lady who represented it counts tokens on a table, following the text of a book with her finger. One might think it is a ledger, but the Latin inscription—which we have placed in the epigraph—that documents the scene refutes this hypothesis.

What is accepted in the category of numbers has evolved over the course of history. The Greeks separated counting numbers from measuring magnitudes. It was not until the seventeenth century, with John Wallis (1616–1703), that a complete parallel was made between numbers and magnitudes. In the article "Number" of the *Encyclopédie ou Dictionnaire raisonné des sciences, des arts et des métiers* (vol. 11, p. 202), there are two definitions of number: a *common* definition—that is to say for the common man—which is the Euclidean definition (Euclid, 1956), namely collection or *assembly of units or things of the same kind*. This is the number to quantify a collection: an academic definition of number, which refers to Newton (1643–1727), which is more about magnitudes: *abstract ratio of one quantity to another of the same kind, which is taken as a unit*. There are three kinds of numbers:

- integers (natural),
- broken (fractions),
- mute (incommensurable or irrational).

It can be noted that the vocabulary used for the third class remains that, devaluing, assigned to it by the ancients, although already denounced by Stevin (*circa* 1548–1620). The problem with these mute numbers is that they touch infinity. They are finite entities (like π or $\sqrt{2}$) which correspond to unlimited sequences of numbers.

It is also noticeable that the two definitions oppose each other by the ontological status of the number: in the common man, it is a collection of entities, in the academic, it is a simple relation, like the relation of parallelism.

When the German Mathematical Society asked David Hilbert (1862–1943) for a report on numbers in 1897, he began his preface in these terms:

> Number theory is one of the oldest branches of mathematics and the human mind became aware very early of some quite deep properties of the natural numbers. Nevertheless, its

status as an independent and systematic science is entirely an achievement of modern times. (Hilbert, 1998, p. vii)

To provide addition and multiplication with inverse operations on all pairs of natural numbers (or almost, the case of zero cannot be solved), it was necessary to create new families of numbers, and this continued to allow every number to have square roots, cubic..., then for every measure of magnitude to be expressed by a number. Pythagorean thought is still evident in the nineteenth century among some mathematicians like Kronecker (1823–1891) who asserts that "God made the whole number, the rest is the work of man". What is affirmed here is that it is impossible to describe all the rules that allow the production of numbers and to describe all of their properties, and that one must start from natural integers. The integers are therefore natural in the sense that they preexist independently of any thought. If the philosopher Berkeley (1685–1753) already posits that the number is entirely the work of the mind, it is necessary to wait for Gauss (1777–1885) to affirm that as a result, it is possible to describe *a priori* all the rules that govern them: constructions, relations and manipulations by operations. It is the next generation that will propose the axioms and rules specific to the construction of integers, called natural, then rational numbers, real, complex, quaternions transforming the aphorism of the *Timaeus* by Plato "God arithmetises" into "Humanity arithmetises".

Scientific and Pedagogical Aspects

Classifications

From the age of 3, arranging, sorting and classifying are important activities, justified by the importance of this type of activities in all science, as highlighted by Henri Poincaré (1854–1912) in his philosophical treatise *The Value of Science* (1905) :

> What is science? [...] it is before all a classification, a manner of bringing together facts which appearances separate, though they were bound together by some natural and hidden kinship. Science, in other words, is a system of relations. (Poincaré, 1913, pp. 349–350)

Any classification requires a prior operation of abstraction "which essentially consists in removing by thought in the thing studied, the particular characters it presents and which would prevent, if we took them into account, from comparing it with other similar things" (Laisant, 1907, p. 16).

However, deciding what should be eliminated and what should be retained among the characteristics of objects is a matter of judgement, and therefore necessarily contextual, even for mathematical objects like numbers. It is a matter of attributing a property or its opposite, which is Plato's preferred mode of categorisation, called *diæresis*. Understanding what a number is for the ancient Greeks is a prerequisite for understanding their classifications.

What Is a Number in Ancient Greece?

In ancient Greece, the number is linked to plurality—which opposes unity. It represents the quantity of discrete entities in a finite collection of elements. It is the common property shared by all collections to have the same answer to the question: exactly how many elements are in this collection? As it is difficult to talk about a collection when it is reduced to one, or even two elements, it is from three that we start to talk about collections. However, the Greeks had three grammatical numbers, the singular or unit, the dual or pair of units and the plural which therefore started at three units. Thus for some, 2 was not a number, but the generator of even numbers, parity being the first of the properties attached to numbers. We will consider it as a number. On the other hand, 1 is not a number, since it is not plurality.

For the Pythagoreans, the world was created according to the system of numbers. They represented these numbers as assemblies of thick points, drawn by a stick on sand or pebbles and classified them according to the geometric figures that could be obtained from these representations.

The definition given by Boethius (1983, p. 76) is twofold: "A number is a collection of unities, or a big mass of quantity issuing from unities". The units align or stack to make a number. Thus, the unit is naturally placed at the head of the sequence of numbers and of any series of figured numbers (Fig. 4.1).

Of course, the same number can be represented in several figures, and therefore have several names. In particular, the numbers are all linear, since they can always be aligned in a single row. Some numbers can be organised into an equilateral triangle, these are the triangular numbers, like 3, 6, 10, others in a square like 4, 9, in a cube like 8, 27. Each form determines a series of numbers having a common character. These characters express some relationships between the number and its parts. Thus $6 = 1 + 2 + 3$ and $6 = 2 \times 3$.

The unit participates in all forms, since it is the head. Thus 6 is linear, oblong and triangular as shown in Fig. 4.2.

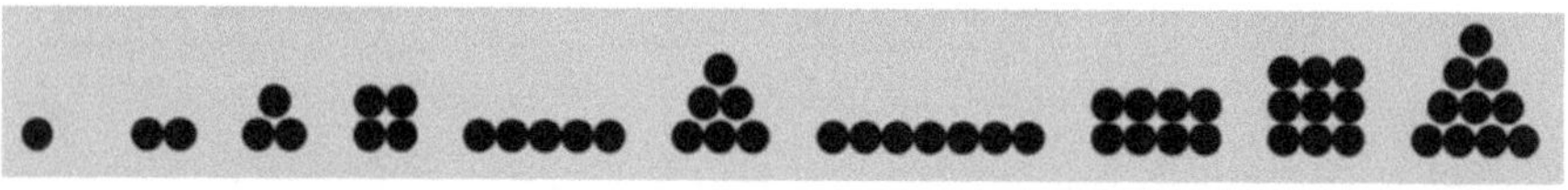

Fig. 4.1 Some figurative representations of the unit and numbers from 2 to 10

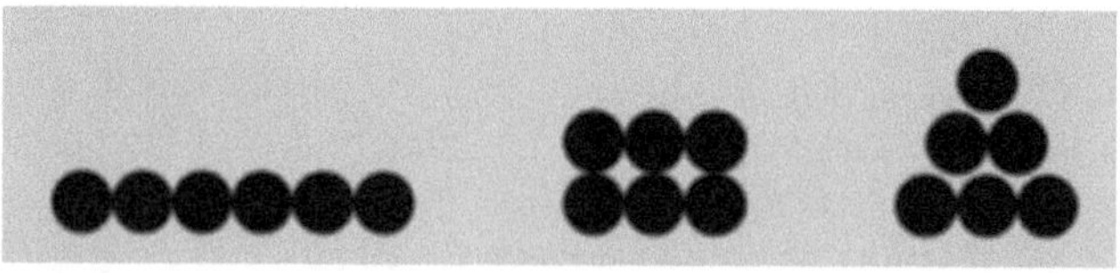

Fig. 4.2 Three figurative representations of the number 6

Properties of Numbers

Parity

From the age of 8, if you ask a pupil—or a future primary school teacher—how to recognise that a number is even, the immediate answer is that this number ends with 0, 2, 4, 6 or 8. This is a property related to the Hindu-Arabic number system. The definition given by Boethius is quite different, and is based on the figurative representation mode: "An even number is one that can be divided into two equal parts, without a unit falling in the middle". Thus, according to Fig. 4.3, 6 is even, and 7 is odd. It should be noted that the definition decomposes into parts and not into numbers, which allows 2 to be included in the classification, as decomposable into two parts of one unit each.

In fact, there is no need to know the name of the number to classify it as even or odd. Therefore, with this approach, it is possible to introduce from the age of 3, the concept of a collection—finite—even or odd, regardless of knowing the quantity represented. One can even easily establish the parity table of the addition of two disjoint collections:

+	Even	Odd
Even	Even	Odd
Odd	Odd	Even

Primality of a Number

A number is *prime* if it is not rectangular, that is, if there is no figurative representation in a rectangle (which can be square for 4). It is noted here that the unit not being a number, there is no need to ensure its exclusion from prime numbers. We also see here that 2 is special, as it is the only prime number that is even. A number that is not prime is said to be *composite*.

The rectangular representation of certain numbers leads to a conception of the multiplication of integers that is not just an abbreviation of addition, but a new operator that is the copy of a number in several copies that are stacked on top of each other. For example, the Fig. 4.4 represents the number twelve as a collection of units (a), as duplication of six, taken as the base of the rectangle (b), as triplication of four, taken as the base of the rectangle (c), as quadruplication of three, taken as

Fig. 4.3 Even or odd? (**a**) 2 is even. (**b**) 6 is even. (**c**) 7 is odd

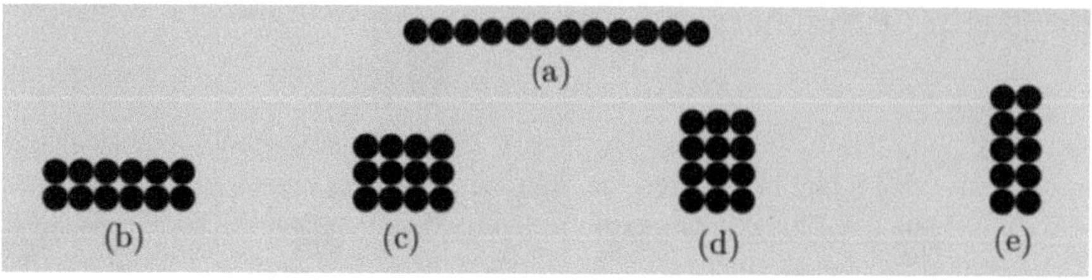

Fig. 4.4 The linear and rectangular figures of twelve and their meanings. (**a**) 12 units aligned. (**b**) 2 times 6. (**c**) 3 times 4. (**d**) 4 times 3. (**e**) 6 times 2

Table 4.1 Table of manyfolde following Recorde (1557, 1582)

Natural numbers	1	2	3	4	5	6	7	8	9	10	11	12	13	14	...
Double	2	4	6	8	10	12	14	16	18	20	22	24	26	28	...
Triple	3	6	9	12	15	18	21	24	27	30	33	36	39	42	...
Fourfold	4	8	12	16	20	24	28	32	36	40	44	48	52	56	...
Fivefold	5	10	15	20	25	30	35	40	45	50	55	60	65	70	...
Sixfold	6	12	18	24	30	36	42	48	54	60	66	72	78	84	...
...															...

the base of the rectangle (d), as sixplication of two, taken as the base of the rectangle (e).

A question that naturally arises is "why do we call *prime* a prime number?". We have seen that the unit is not a number, but it was the first of the numbers, the one that is placed first, and which generates the numbers one after the other (the flow of the unit). This is only from the additive point of view. The multiplicative point of view allows us to answer this question.

Let's observe a table of series of multiples as it appears in Boethius and in Table 4.1. A number is prime if it does not appear in any other series of multiples than the one of which it is the head. To verify this, it is enough to have the list of multiples of the numbers which are inferior to it. Thus 2, 3, and 5 are prime, 4 and 6 are composed, as they both appear in the series of doubles. Prime numbers are indeed the elementary multiplicative bricks of numbers, as stated by the fundamental theorem of arithmetic: "every integer can be uniquely decomposed (up to order) into a finite product of prime numbers (not necessarily distinct)".

The Relationships Between Numbers

In everyday language, a fraction—or a broken number—is a part of a whole, it therefore cannot exceed one. Pupils aged 6 to 9 have difficulty accepting that a fraction can express something greater than one. In English, we distinguish between *proper fractions* and improper fractions. It's one thing to admit four thirds ($\frac{4}{3}$) as four times a third, or $4 \times \frac{1}{3}$, it's another to consider the result obtained as a fractional number. Until the sixteenth century, an improper fraction was thought as a whole followed by a proper fraction. Thus four times a third will be written $1\frac{1}{3}$ and said one and a third, which we still say in English. We will use this notation to give a fractional representation of number ratios. These expressions composed of a whole number and a *proper fraction* are called *mixed numbers*.

Fractions were noted without a bar by the Hindus and Arabs. The Hindus also used vertical notation for broken numbers. It was Leonardo of Pisa, following Maghrebian mathematicians, who encouraged the notation with a horizontal separation bar $\frac{4}{3}$, which could become, for practical printing problems, a slash bar $4/3$.

In natural language, each fraction has a name, formed with the cardinal of the denominator and the ordinal of the numerator, but no word in the lexicon designates the series of all the fractions that represent the same quantity, that is to say the associated rational. This is not the case with the Greeks who will give a name to the series to designate the quantity represented, this name exactly describing the broken number with the smallest denominator and numerator, as a generating element of the class. We can say that each fraction has, for the Ancients, a surname—the name of the series, and a first name, its expression.

This approach allows to clearly differentiate expression of a number and the number itself, which unfortunately is not always acquired by pupils, even those intending to teach in schools.

For the Ancients, a fraction represented a certain multiplicative ratio, translated into Latin by *proportio* then in old and middle French or English by *proportion*. We prefer to adopt the most neutral term, that of *ratio* which we will note (Number1;Number2) and which reads "ratio of Number1 to Number2".

It is very important to note that like many usual relational properties involving opposites, like parent/child, master/disciple, older/younger..., relations are oriented. Thus each number, according to its place in the pair, plays a particular role, and these roles are exchanged when the order is reversed.

The organisation of number ratios into series is the culmination of a classificatory approach specific to the Greeks, which has been transmitted at least until the Renaissance in academic circles. It is this approach that we will now present to arrive at the nominalisation of rational numbers.

Classification According to Additive Comparison

To classify number ratios into categories, we first follow the additive point of view.

There are three kinds of additive ratios between numbers (Fig. 4.5a), which the Greeks obtain by *diæresis* in two stages following the scheme of Fig. 4.5b. Let the two numbers be either equal or unequal.

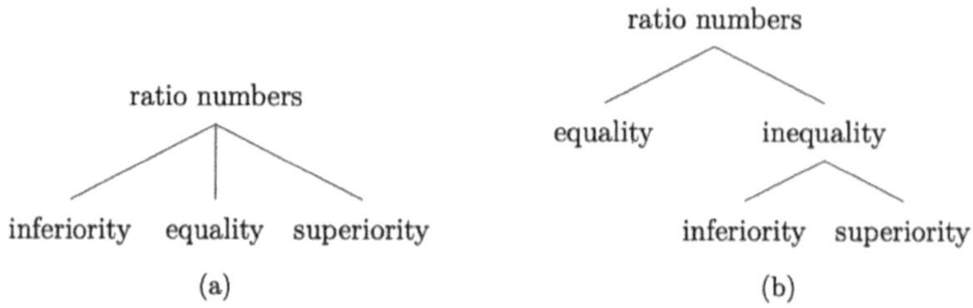

Fig. 4.5 Two reasoning schemes for classifying. (**a**) Case-base reasoning. (**b**) Reasoning based on diæresis

In the first case, we are dealing with a ratio of equality, such as (1;1), (2;2), (3;3), ... This category contains only one series of number ratios, that of the unit which generates this series in one-to-one correspondence (bijection) with the natural numbers.

In the other branch of the alternative are all the other number ratios. We can then distinguish two types of inequality ratios: either Number1 is contained in Number2 (as a collection of units), and we will say we have a ratio of inferiority, or Number1 contains Number2 and it is a ratio of superiority.

This first step is within the reach of the youngest pupils (from 7 years old), for the part of the numerical rhyme mastered. It also provides a beautiful example of nested conditional reasoning. This is the reasoning by *diæresis* found in the authors studied, the Frenchman Nicolas Chuquet (*circa* 1445/55–*circa* 1488/1500) in Chuquet (1880) and the Welshman Robert Recorde (*circa* 1512–1558) in Recorde (1557, 1582).

There is a one-to-one correspondence between the two categories: (Number1;Number2) is in one category if and only if (Number2;Number1) is in the other. It is therefore natural, and economical, to continue the division for one of the categories, and to deduce the division on the other. This is what the Greeks did, working on the category of superiority ratios.

From a pedagogical point of view, this is a very good thing, since it makes the quantities greater than the unit work, those which are less easily admitted as fractions.

First Subdivision: Multiple/Non Multiple

The additive point of view being qualitatively exhausted, we move on to the multiplicative point of view. The first relation that then imposes itself between two natural numbers is the multiple/divisor ratio. We have already worked on these types of ratios between a number and one of its parts. It is now a question of seeing if Number1 is a multiple of Number2, that is to say if in terms of figurative representation Number1 is rectangular with base Number2. The number of times

Table 4.2 Table of multiple ratios

Double ratio	**(2;1)**	(4;2)	(6;3)	(8;4)	(10;5)	(12;6)	(14;7)	(16;8)	...
Triple ratio	**(3;1)**	(6;2)	(9;3)	(12;4)	(15;5)	(18;6)	(21;7)	(24;8)	...
Quadriple ratio	**(4;1)**	(8;2)	(12;3)	(16;4)	(20;5)	(24;6)	(28;7)	(32;8)	...
⋮	⋮	⋮	⋮	⋮	⋮	⋮	⋮	⋮	⋱

Number2 is stacked to represent Number1 is called *quotient*, from the Latin adverb *quotiens* [how many times]. Each quotient corresponds to a ratio name (in the following, we adopt Recorde's terminology): double, triple, quadriple, quintiple, sextiple, septiple, octiple for the quotients 2, 3, 4, 5, 6, 7, 8, and more generally n-tiples for the quotient n.

It is therefore possible to classify all multiple ratios in series, each bearing the name of the associated quotient, as shown in Table 4.2, built on Table 4.1.

Multiple inferiority ratios, of the form (Number1;*mult*×Number1) will be said under *mult*-tiple. (7;21) is a *sub-triple* ratio, and not *third*, which we would naturally say. It remains therefore to put in order the other number ratios.

Classification of Non-multiple Ratios

Before starting, several remarks are necessary. First, each series already constituted is an image of the series of natural numbers generated by the unit. A ratio of two numbers in its series is obtained by multiplying each term of the ratio by the number associated with its rank. Thus (18;6) is the sixth element of the triple ratio series, it is obtained by multiplying by 6 the two leading members, which can be written as follows: (18;6) = 6 times (3;1). But if the ratios are not multiples, then there will be a remainder.

We must now remember that for the Ancients, 1 is not a number, and that a multiple of a number is, as in the current common language, several times the number. Thus, the Euclidean division of Number1 by Number2 produces a non-zero quotient (*mult*) and a non-zero remainder (*remainder*) strictly less than Number2. We can write for (Number1;Number2):

$$\text{Number1} = mult \times \text{Number2} + \text{remainder},$$
$$mult \geq 1,\ 0 < \text{remainder} < \text{Number2}$$

What then happens for the number ratio:

λ times (Number1;Number2) = ($\lambda \times$Number1; $\lambda \times$Number2), λ being any natural number. The Euclidean division can be written:

$\lambda \times$Number1= $mult \times (\lambda \times$Number2) + $\lambda \times$remainder,
$mult \geq 1$, 0< remainder <Number2

The quotient is the same, but the remainder has undergone the common expansion of Number1 and Number2, which accurately reflects the fact that any common divisor to Number1 and Number2 is a common divisor to Number2 and *remainder*, that is to say that (Number1;Number2) and ($\lambda \times$Number1;$\lambda \times$Number2) are in the same series, whose head is obtained by reducing Number1 and Number2 to two numbers without common divisors, that is to say into two numbers *prime to each other*.

This is a property of ratios, analogous to the property of numbers to be prime, since it discriminates, among non-multiple ratios, the heads of series from the others.

Reducible/Irreducible Ratio
A ratio is *irreducible* if the two numbers are not multiples of a third number, otherwise, it is said *reducible*. In terms of figurate numbers, a ratio is reducible if these two numbers can be rectangularised with a common base. This common base constitutes a reduction factor of the ratio.

This new subdivision therefore allows to determine all the series of ratios, by associating with each ratio of irreducible numbers a name, which the series will inherit as is commonly the case to designate the whole by a part, as in everyday language *day* for the *nychthemeron*, the day/night period of 24 hours.

Before assigning a name to each series, the Greeks further refine the class of irreducible ratios according to whether respectively the quotient and the remainder are equal or different from 1 following Tables 4.3 and 4.4.

Thus (13;3) is a manyfolde *superparticulare ratio*, because 13=4$\times$3+1 associated with the mixed number 4 1/3; and (3;13) is a submanyfolde *subparticulare* by duality.

Table 4.3 Name of the class of a non-multiple irreducible ratio of superiority

Number1=$mult \times$ Number2 + remainder (Number2 <Number1)	Remainder=1	$2 \leq$ remainder < Number2
$mult = 1$	Superparticulare	Superpartiente
$mult \geq 2$	Manyfolde Superparticulare	Manyfolde superpartiente

Table 4.4 Assignment of the class of a non-multiple irreducible ratio of inferiority

Number2=$mult \times$Number1 + remainder (Number1 <Number2)	Remainder=1	$2 \leq$ remainder < Nombre2
$mult = 1$	Subparticulare	Subpartiente
$mult \geq 2$	Submanyfolde subparticulare	Submanyfolde subpartiente

Thus (13;5) is a manyfolde *superpartiente ratio*, because 13 breaks down into a sum of several 5s with a remainder greater than 1, and, by duality, (5;13) is a submanyfolde *subpartiente ratio*. As spelling did not stabilise until the nineteenth century, to accompany compulsory schooling, we also find *parciens* instead of *partiens*. We have opted for *partiens* closer to *part*. We find *sur* sometimes instead of *super*.

Name of Each Series of Non-multiple Superiority Ratio
The name of a ratio should allow to discriminate its irreducible ratio from other irreducible ratios. By examining once again the Euclidean division of Number1 by Number2, the ratio (Number1;Number2) is entirely determined by the triplet (Number2;*mult*;*remainder*).

mult is expressed as the multiple ratio. Number2 is the unit by which Number1 is reported. Number2 is an enlargement of the unit. Number2 also allows to divide any segment into Number2 equal parts, which explains the use of the ordinal by the Ancients: the Number2th part of the unit (Fig. 4.6). *Remainder* tells the quantity of simple units remaining, which are as many Number2th parts relative to Number2.

Table 4.5 gives the names of the values. We have signified the impossibility by a line and the absence of a term by the absence of a sign in the box. Not all combinations of a term in each series directly give the name of the series. Some can describe a ratio of reducible numbers. Thus, the elements of the same column to say Number2 and *remainder* do not never form the name of a ratio, since then the ratio (Number1;Number2) is reducible. It must then be reduced to find the name of the ratio.

The terms are derived from Latin. Thus *sesqui* first meant *and a half more*, then *an nth and more*. Medieval and Renaissance scholars communicated among themselves in Latin, which explains why these denominations have remained, since there was no equivalent in the vernacular language to express the quantity without describing its representation.

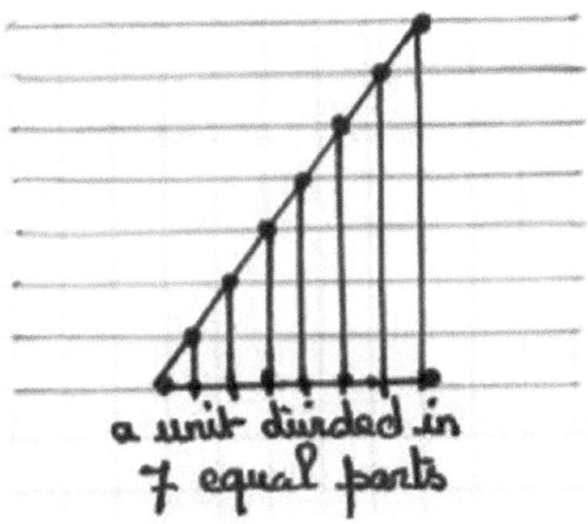

Fig. 4.6 Multiply to divide better

Table 4.5 Naming of superiority ratios according to the irreducible ratio (Number1;Number2) and the Euclidean division Number1 =*mult*× Number2 + *remainder*

To say	0	1	2	3	4	5	6	7	...
Number2	–	–	Altera	Tertia	Quarta	Quinta	Sexta	Septima	...
(translated)			Half	Third	Fourth	Fifth	Sixth	Seventh	...
mult	–		Double	Triple	Quadriple	Quintiple	Sextiple	Septiple	...
remainder		sesqui	Bi-	Tri-	Quadru-	Quintu-	Sextu-	Septu-	...

Table 4.6 The series of the sesquialter ratio

The natural numbers	1	2	3	4	5	6	7	8	9	...
The sesquialter ratio	(3;2)	(6;4)	(9;6)	(12;8)	(15;10)	(18;12)	(21;14)	(24;16)	(27;18)	...

In arithmetic treatises, the practical part on fractions and their calculations precedes this more theoretical part, when this latter exists, which is not the case in practical arithmetic. One of the first treatises written in French and still accessible, that of Nicolas Chuquet (1880) is divided into three parts. The first part, divided into four chapters, deals with the representations and elementary operations of whole numbers (first chapter), of broken numbers or fractions (chapter 2). In chapter 3 appears the academic study of numbers: the *progressions* in connection with the figured numbers, properties of numbers such as parity, primality, perfection, followed by the study of number ratios—on which we have relied for the drafting of this chapter. The fourth part concerns the rules of three. Similarly, the first author in English, the Welsh Robert Recorde only deals with this part in his second book of arithmetic (Recorde, 1557), seventeen years after the first part, intended for beginners (Recorde, 1582). On the other hand, in 1725, when Johann Joseph Fux publishes in Latin a treatise on music theory, the first chapter is devoted to the presentation of number ratios (Fux & Miller Schulte, 2017).

Thus, following the indications of Table 4.5 for the types of a number ratio (Number1;Number2), satisfying:

Number1 = $mult\times$Number2 + $remainder$

herefore represented by the mixed number $mult\ remainder/Number2$.

- **superparticulare** sesqui[Number2]

  Examples: **(3;2)**: we get 3 = (1 ×) 2 + 1, then Number2 = 2, (*mult*= 1), remainder = 1, therefore associated with 1 1/2.

  The ratio is *sesquialtera*, the series of ratios thus named is given by Table 4.6.

  (6;5) : 6 = (1×) 5 + 1, Number2 = 5, (*mult*= 1), remainder = 1, associated with 1 1/5.

  It is therefore a *sesquiquinta* ratio.
- **superpartiens** super[remainder]partiens[Nombre2]s

  The plural mark signals the fact that one must add to Number2 several equal parts of the unit to obtain Number1. For example (5;3) is a ratio *superbipartiens tertias*, because $5 = 1 \times 3 + 2$, associated with 1 2/3 and (11;7) is a ratio *superquadrupartiens septimas*, because $11 = 7 + 4$, associated with 1 4/7.

For the next two types, it is necessary to signify the multiplicity of Number2 in Number1. Depending on the value of the remainder, we have

- **manyfolde superparticulare** [*mult*]sesqui[Nombre2]

  For example (5;2) is a ratio *double sesquialter* because $5 = 2 \times 2 + 1$ associated with 2 1/2, and (13;4) is a ratio *triple sesquiquarta*, because $13 = 3 \times 4 + 1$ associated with 3 1/4.
- **multiple superpartiens** [*mult*]super[remainder]*partiens*[Nombre2]**s**

  For example (8;3) is a ratio *double superbipartiens tertias*, because $8 = 2 \times 3 + 2$ associated with 2 2/3, and (31;7)] is a ratio *quadriple supertripartiente septimas*, because $31 = 4 \times 7 + 3$ associated with 4 3/7.

Names of Non-multiple Inferiority Ratios

It is enough to replace *multiple* with *sub-multiple* and *super* with *sub*. Thus, the ratio (13;5) is a *double superbipartiente quintas* ratio and (5;13) is a *subdouble subbipartiente quintas* ratio.

Thus concludes the classification of number ratios into series defining each a unique quantity, which we call rational number. Understanding these compound expressions is one of the most interesting coding/decoding and lexical exercises.

What the Ratio Name Says

To speak of Louis XIV or to speak of the Sun King is to refer to the same person. However, these two expressions do not say the same thing. The first expression places him within a dynasty and a kingdom, the second describes a way of being.

The name of the ratio of numbers gives the code to locate it on a scale graduated by natural numbers. Any ratio having this name, whether it is reducible or not, allows to locate on the straight line graduated by the unit the set of ratios of the same name from the Euclidean division of Number1 by Number2. The multiplicity *mult* gives the natural number *mult* that approaches it best by default, the *remainder* allows to locate it on the segment bounded by *mult* and $mult + 1$ thanks to a suitable guide-donkey like the one in (Fig. 4.6). The division is first made on a segment constructed according to the desired division, then projected onto the segment to be divided.

Figure 4.7 shows how to locate the ratios *sesquialter* (A) and *supertripartiente septimas* (B) from a guide-donkey drawn above the numerical half-line. It is a matter of drawing enough parallels to the half-line to exceed by one unit Number2. We have drawn 13, which allows to decompose any unit segment into at most 12 equal parts. To divide a unit segment $[u, u + 1]$ into k equal parts, it is enough to raise a perpendicular at the point u up to the first unnumbered line, and a perpendicular at the point $u + 1$ up to the numbered line k. Between the two points thus defined, a segment is drawn, which is an image of the unit segment. The $k + 1$ intersection

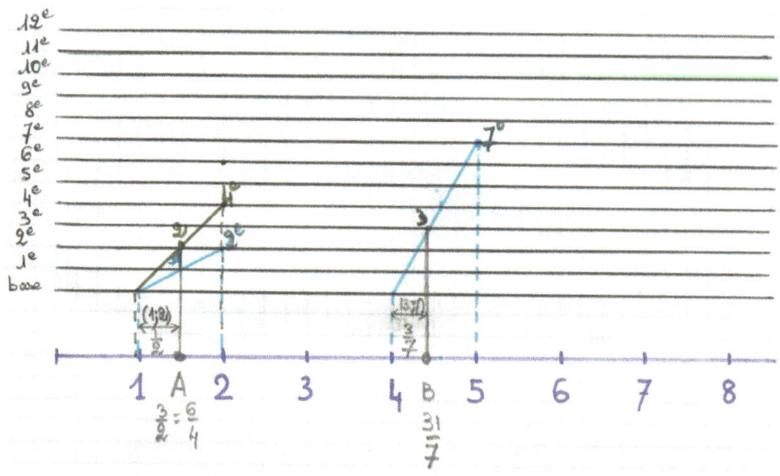

Fig. 4.7 Locate a ratio of numbers or a fraction on a graduated rule to the unit, with a guide-donkey

points of this image with the labeled lines base, 1 to *k* divide this image into *k* equal parts, their orthogonal projection to the initial segment divides the latter into as many equal parts.

The ratio *sesquialter*, associated with the mixed number 1 1/2 is located in the middle of the segment with ends 1 and 2. (3;2) and (6;4) are two of its representatives. The ratio gives the name of the rational and the guide-donkey allows to locate it on the line.

A ratio *quadriple supertripartiente septimas*, associated with the mixed number 4 3/7, has for irreducible representative $(4 \times 7 + 3; 7)$, or (31;7).

Implementation in the Classroom

This sequence was worked on in 2016–2017 with the teachers of the last two years of elementary school, pupils aged 9 to 11 from the Orméteau school. Between October and February, conferences by the author were scheduled with the teachers and were reworked to be done independently by the teachers and proposed for the training through research of future teachers (wishing to become primary school teachers). In particular, we try to show that primary school mathematics cannot ignore the historical aspect, since it concerns its foundations, which the Ancients laid down, before the Moderns formalised them to allow their generalisation. The current mathematical language borrows from the Ancients a vocabulary often passed into everyday language. The latter often retains the original meaning, while mathematics has adapted it to the needs of the discipline. We have often noticed in our teachings, from school to university that many blockages in mathematics come from this gap between the old meanings rooted in usage and the new institutional meaning.

It was decided to proceed in three periods: a first period focused on numbers themselves, a second on the multiple/divisor relationships and the third on number ratios. The pupils' abilities to classify, represent and interpret mathematical objects (here numbers) and their properties from different points of view, which are necessary foundations for the development of critical thinking of future citizens and scientific work, are systematically developed. This approach also allows to value in the pupil the habit of figuratively representing objects by dealing with small examples by hand, not only to make conjectures, but also to develop reasoning approaches, which is sorely lacking in undergraduate pupils.

Pupils work in groups of two or four depending on the class configurations. We detail below the different sequences noting what was done in the cited experiment.

Between September and November: Representation of Numbers

Materials

- tokens of the same size in the form of a disc or square. If possible a hundred per group,
- two tables to fill in: one on the rules of parity and the other to note the triangular, square, rectangular numbers, at least the first ten.

+	Even	Odd
Even		
Odd		

Order in the series	3rd	4th	5th	6th	7th	8th	9th	10th	11th	12th	...
Square numbers											
Triangular numbers											

First Session: Historical Aspect and Work on Parity

The teacher explains why the Pythagoreans were interested in numbers in themselves and how they represented them. It was a work of philosophers, all adults and chosen by the master. Then the tokens and the tables are distributed. Indeed, this representation, which has certainly been used already in previous classes, allows to illustrate properties of numbers, and to deduce calculation rules.

Parity The pupils are invited to explain how to define parity for a set of tokens. The proposals are noted on the board, then Boethius's definition is given. It is useful also to highlight, if it has not emerged, the notion of pairing tokens two by two and the fact that one does not need to know the name of the number to say whether it is even or odd.

Then, the pupils must explain how to fill in the parity table for addition, trying to clearly differentiate between illustration by an example and general explanation, which consists of making a reasoning by cases according to the number of unpaired tokens and the groupings that can be made. The goal is for the pupils to have understood that it is the odd numbers that decide the result.

For the most advanced pupils, the question can be extended to a sum of several terms, and arrive at the following statement: *a sum of (integer) numbers is odd if there is an odd number of odd terms*. This statement uses the term number twice, the first time as elements of the set, and a second time as the quantity of a set of terms. Even stronger, a similar table can be done with multiplication.

Second Session: Square Numbers and Triangular Numbers

Relying for example on Fig. 4.1, the teacher defines what a triangular number and a square number are. Given the special place of the unit among the Pythagoreans, it is necessary to specify that the unit is the generator of all types of regular figures, in particular, it is a polygonal number of any order, it is therefore the first of triangular, square, pentagonal numbers... For square numbers, one can introduce the vocabulary of the side of the square in connection with the root of the square number, and the area of the square in connection with the square number. For triangular numbers, it is necessary to point out that it is not just any type of triangle, but only the one whose three sides are the same length, that is the equilateral triangle.

The pupils can start without any other instruction to fill in the table. The first four triangular numbers and the first three square numbers are given by Fig. 4.1. Some will start by trying for each number, whether it is triangulable or squarable and find that even in a group, it's a bit of a long job. Some will have tried to look at how we move from a figure with side 2 to the figure with side 3. For that purpose, the table to be filled had to get enough columns to convince pupils to scrutinize Fig. 4.1. for finding by themselves a constructive rule. This is therefore a classic exercise in constructing logical sequences.

The nth square number is constructed by adding to the number $n + (n - 1)$ the $(n - 1)$th square number, or by adding to the $(n - 1)$th square number twice the number $n - 1$, once in line and once in column, and the unit, which illustrates the remarkable identity $(n + 1)^2 = n^2 + 2n + 1$ (Fig. 4.9). The square numbers are the well-known numbers 1, 4, 9, 16, 25, 36, 49, 64, 81, 100, 121...

The nth triangular number is constructed by adding to the number n (the length of its side) the $(n - 1)$th triangular number starting from 1 (Fig. 4.8). Thus we find 1, 3, 6, 10, 15, 21, 28, 36, 45, 55, 66... We will notice that there is an alternation between two odd numbers and two even numbers.

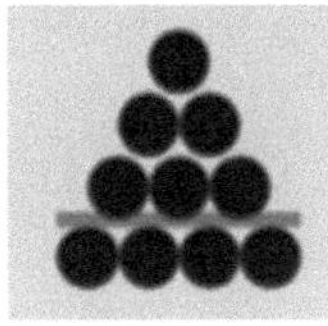

Fig. 4.8 Construction of the 4th triangular number: 10, from its predecessor

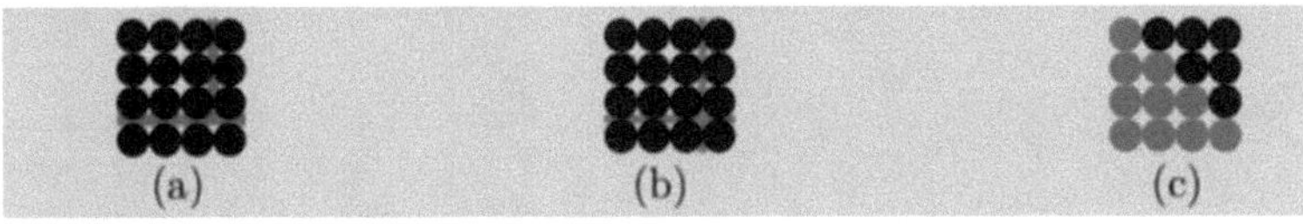

Fig. 4.9 Constructions of the 4th square number: 16, from its predecessor or from triangular number with same order and its predecessor. (**a**) $4^2 = 3^2 + 4 + 3$. (**b**) $4^2 = 3^2 + 2 \times 3 + 1$. (**c**) $4^2 = 10 + 6$

We can also notice that two successive triangular numbers give a square by addition. The fourth square is constructed by adding the fourth and third triangular numbers (Fig. 4.9).

Between November and January: First Relationships Between Two Numbers, Multiples/Divisors

The objective of this work period is, on the one hand, to establish tables of multiples to compensate for the lack of ease in the use of multiplication tables and to get pupils to work on them differently, and, on the other hand, to express properties concerning divisibility from multiple points of view: with figurative numbers like the Pythagoreans and from the tables constructed. The manipulative material forces us to consider only fairly small numbers (less than a hundred), but the beginning of the third session, which consists of comparing two integers, can be the pretext for revising very large numbers.

Materials
For the first session: tokens; sets of dominoes without the "zeros" then a game of pairs of numbers where the first member can be distinguished from the second (indeed, (2;3) is different from (3;2) by the order of reading, which is not possible with dominoes), a small box, a large box and two medium ones that can go in the large box or a small envelope, two medium ones and a large one to store the reports.

For the second session: at least one support for the tables of multiples/divisors. In the school where we are experimenting, small 10×10 tables of individual numbers from 1 to 99 have been chosen (an A4 sheet can contain about ten like those in Fig. 4.10 or collective A3 format tables to display in the classroom. The bias—or

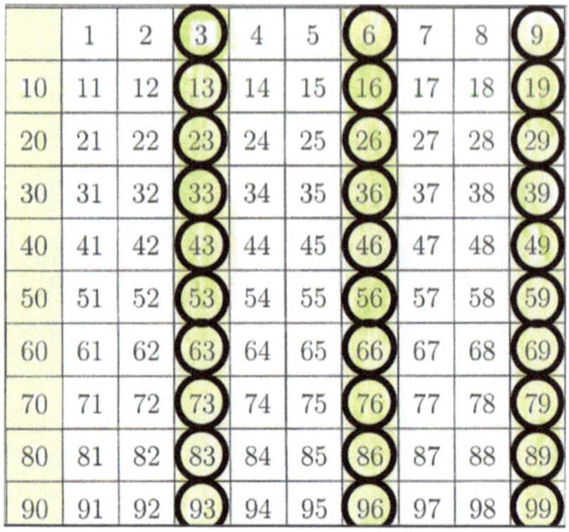

Fig. 4.10 Multiples of 3 as those of 2

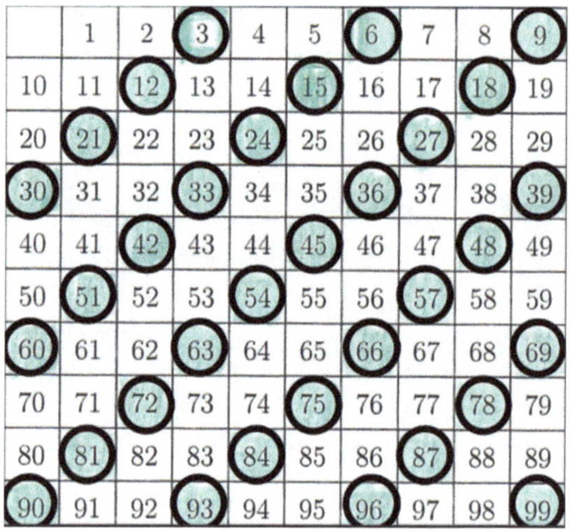

Fig. 4.11 The correct table of multiples of 3

advantage—is that it is linked to the decimal system: multiples of 2, 5 and 10 draw vertical bands (Fig. 4.11).

To vary the supports, in order to highlight what depends on the support and what is specific to the series of multiples, one can vary the sizes of the tables, or use Ulam tables (Fig. 4.12), a table of multiples in the form of a timeline, like the one described in Fig. 4.13.

Finally, several small boxes and one intermediate box between small (for given multiplicity ratios) and medium (for multiple ratios) should be available.

The first session is a reflection session on the classifications and supports used. For the experiment, it was not done in class, but tested in workshops with pupils of the same level, supervised by future primary school teachers.

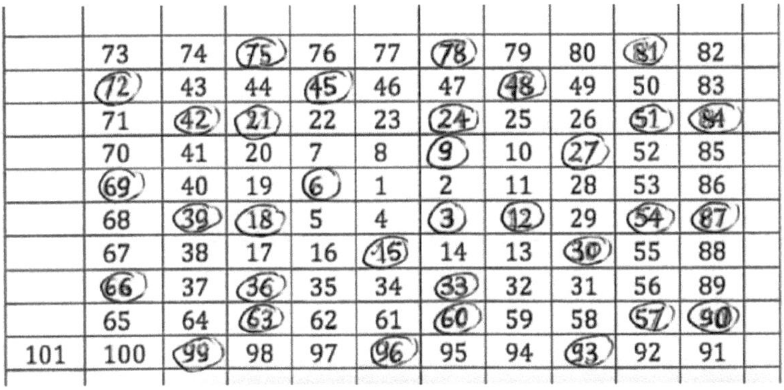

	73	74	75	76	77	78	79	80	81	82	
	72	43	44	45	46	47	48	49	50	83	
	71	42	21	22	23	24	25	26	51	84	
	70	41	20	7	8	9	10	27	52	85	
	69	40	19	6	1	2	11	28	53	86	
	68	39	18	5	4	3	12	29	54	87	
	67	38	17	16	15	14	13	30	55	88	
	66	37	36	35	34	33	32	31	56	89	
	65	64	63	62	61	60	59	58	57	90	
101	100	99	98	97	96	95	94	93	92	91	

Fig. 4.12 Ulam spiral and the multiples of 3

First Session: Classification of Ratios I, Additive Comparison

In this session, the teacher begins by reminding the pupils that for the Pythagoreans, numbers are beings that have properties, they can be even or odd, triangular, square... They also have relationships with each other. When a relationship is established between two elements, there is always one element that is taken as a reference and the other that is compared. To say that Jean is smaller than Claire, is to take Claire as a reference element, and measure Jean in relation to Claire. To say that Claire is taller than Jean, is to take Jean as a reference and compare Claire to Jean. We therefore have two expressions which describe the same qualitative fact: the existence of a difference in size, but from two different points of view. For numbers, it will be the same thing, the first number of the ratio will be compared to the second. By convention, we write (Number1;Number2) to express the ratio of Number1 to Number2. Explaining that the Greeks make a first sort between being equal and being different may seem superfluous, and one might be tempted to classify directly into three parts. It is however an important step, because equality, which is here an identity between numbers, is the first perfectly constituted class.

We then ask what this sorting means on the dominoes. It is about sorting them according to whether they are doubles or not. This equality class has a name, the other class is made up of the other dominoes. We put the doubles in the small box and the others in the large one. In the spirit of our Pythagoreans, whole numbers being used to represent the universe, we can dwell on this class and ask the question of what number could represent this class. The pupils, now a bit philosophical, find that it is unity, because "two equal numbers are the same and it's like for fractions (2;2), (3;3), (4;4), ... it's the same thing, it's always one" (sic). We label the small box with "equality, 1".

We then ask the pupils if we can sort the remaining dominoes into larger and smaller ratios. This is of course impossible since the dominoes have no reading direction, they can be placed in one direction or the other. We therefore need to change the support. The writing in numbers of pairs of numbers (2;3) and (3;2) are indeed two different pairs. We therefore cut out on squared cardboard pairs (1;1),

1	1	1	1	1	1	1	1	1	1	1	1	1	1	1	1	1	1	1	1	1	1	1	1	1	1	1	1	1	1	1	1	1
2	**2**	2	2	2	2	2	2	2	2	2	2	2	2	2	2	2	2	2	2	2	2	2	2	2	2	2	2	2	2	2	2	2
3	3	**3**	3	3	3	3	3	3	3	3	3	3	3	3	3	3	3	3	3	3	3	3	3	3	3	3	3	3	3	3	3	3
4	**4**	4	**4**	4	4	4	4	4	4	4	4	4	4	4	4	4	4	4	4	4	4	4	4	4	4	4	4	4	4	4	4	4
5	5	5	5	**5**	5	5	5	5	5	5	5	5	5	5	5	5	5	5	5	5	5	5	5	5	5	5	5	5	5	5	5	5
6	**6**	**6**	6	6	**6**	6	6	6	6	6	6	6	6	6	6	6	6	6	6	6	6	6	6	6	6	6	6	6	6	6	6	6
7	7	7	7	7	7	**7**	7	7	7	7	7	7	7	7	7	7	7	7	7	7	7	7	7	7	7	7	7	7	7	7	7	7
8	**8**	8	**8**	8	8	8	**8**	8	8	8	8	8	8	8	8	8	8	8	8	8	8	8	8	8	8	8	8	8	8	8	8	8
9	9	**9**	9	9	9	9	9	**9**	9	9	9	9	9	9	9	9	9	9	9	9	9	9	9	9	9	9	9	9	9	9	9	9
10	**10**	10	10	**10**	10	10	10	10	**10**	10	10	10	10	10	10	10	10	10	10	10	10	10	10	10	10	10	10	10	10	10	10	10
11	11	11	11	11	11	11	11	11	11	**11**	11	11	11	11	11	11	11	11	11	11	11	11	11	11	11	11	11	11	11	11	11	11
12	**12**	**12**	**12**	12	**12**	12	12	12	12	12	**12**	12	12	12	12	12	12	12	12	12	12	12	12	12	12	12	12	12	12	12	12	12
13	13	13	13	13	13	13	13	13	13	13	13	**13**	13	13	13	13	13	13	13	13	13	13	13	13	13	13	13	13	13	13	13	13
14	**14**	14	14	14	14	**14**	14	14	14	14	14	14	**14**	14	14	14	14	14	14	14	14	14	14	14	14	14	14	14	14	14	14	14
15	15	**15**	15	**15**	15	15	15	15	15	15	15	15	15	**15**	15	15	15	15	15	15	15	15	15	15	15	15	15	15	15	15	15	15
16	**16**	16	**16**	16	16	16	**16**	16	16	16	16	16	16	16	**16**	16	16	16	16	16	16	16	16	16	16	16	16	16	16	16	16	16
17	17	17	17	17	17	17	17	17	17	17	17	17	17	17	17	**17**	17	17	17	17	17	17	17	17	17	17	17	17	17	17	17	17
18	**18**	**18**	18	18	**18**	18	18	**18**	18	18	18	18	18	18	18	18	**18**	18	18	18	18	18	18	18	18	18	18	18	18	18	18	18
19	19	19	19	19	19	19	19	19	19	19	19	19	19	19	19	19	19	**19**	19	19	19	19	19	19	19	19	19	19	19	19	19	19
20	**20**	20	**20**	**20**	20	20	20	20	**20**	20	20	20	20	20	20	20	20	20	**20**	20	20	20	20	20	20	20	20	20	20	20	20	20
21	21	**21**	21	21	21	**21**	21	21	21	21	21	21	21	21	21	21	21	21	21	**21**	21	21	21	21	21	21	21	21	21	21	21	21
22	**22**	22	22	22	22	22	22	22	22	**22**	22	22	22	22	22	22	22	22	22	22	**22**	22	22	22	22	22	22	22	22	22	22	22
23	23	23	23	23	23	23	23	23	23	23	23	23	23	23	23	23	23	23	23	23	23	**23**	23	23	23	23	23	23	23	23	23	23
24	**24**	**24**	**24**	24	**24**	24	**24**	24	24	24	**24**	24	24	24	24	24	24	24	24	24	24	24	**24**	24	24	24	24	24	24	24	24	24
25	25	25	25	**25**	25	25	25	25	25	25	25	25	25	25	25	25	25	25	25	25	25	25	25	**25**	25	25	25	25	25	25	25	25
26	**26**	26	26	26	26	26	26	26	26	26	26	**26**	26	26	26	26	26	26	26	26	26	26	26	26	**26**	26	26	26	26	26	26	26
27	27	**27**	27	27	27	27	27	**27**	27	27	27	27	27	27	27	27	27	27	27	27	27	27	27	27	27	**27**	27	27	27	27	27	27
28	**28**	28	**28**	28	28	**28**	28	28	28	28	28	28	**28**	28	28	28	28	28	28	28	28	28	28	28	28	28	**28**	28	28	28	28	28
29	29	29	29	29	29	29	29	29	29	29	29	29	29	29	29	29	29	29	29	29	29	29	29	29	29	29	29	**29**	29	29	29	29
30	**30**	**30**	30	**30**	**30**	30	30	30	**30**	30	30	30	30	**30**	30	30	30	30	30	30	30	30	30	30	30	30	30	30	**30**	30	30	30
31	31	31	31	31	31	31	31	31	31	31	31	31	31	31	31	31	31	31	31	31	31	31	31	31	31	31	31	31	31	**31**	31	31
32	**32**	32	**32**	32	32	32	**32**	32	32	32	32	32	32	32	**32**	32	32	32	32	32	32	32	32	32	32	32	32	32	32	32	**32**	32
33	33	**33**	33	33	33	33	33	33	33	**33**	33	33	33	33	33	33	33	33	33	33	33	33	33	33	33	33	33	33	33	33	33	**33**
34	**34**	34	34	34	34	34	34	34	34	34	34	34	34	34	34	**34**	34	34	34	34	34	34	34	34	34	34	34	34	34	34	34	34
35	35	35	35	**35**	35	**35**	35	35	35	35	35	35	35	35	35	35	35	35	35	35	35	35	35	35	35	35	35	35	35	35	35	35
36	**36**	**36**	**36**	36	**36**	36	36	**36**	36	36	**36**	36	36	36	36	36	**36**	36	36	36	36	36	36	36	36	36	36	36	36	36	36	36
37	37	37	37	37	37	37	37	37	37	37	37	37	37	37	37	37	37	37	37	37	37	37	37	37	37	37	37	37	37	37	37	37
38	**38**	38	38	38	38	38	38	38	38	38	38	38	38	38	38	38	38	**38**	38	38	38	38	38	38	38	38	38	38	38	38	38	38
39	39	**39**	39	39	39	39	39	39	39	39	39	**39**	39	39	39	39	39	39	39	39	39	39	39	39	39	39	39	39	39	39	39	39
40	**40**	40	**40**	**40**	40	40	**40**	40	**40**	40	40	40	40	40	40	40	40	40	**40**	40	40	40	40	40	40	40	40	40	40	40	40	40
41	41	41	41	41	41	41	41	41	41	41	41	41	41	41	41	41	41	41	41	41	41	41	41	41	41	41	41	41	41	41	41	41
42	**42**	**42**	42	42	**42**	**42**	42	42	42	42	42	42	**42**	42	42	42	42	42	42	**42**	42	42	42	42	42	42	42	42	42	42	42	42
43	43	43	43	43	43	43	43	43	43	43	43	43	43	43	43	43	43	43	43	43	43	43	43	43	43	43	43	43	43	43	43	43
44	**44**	44	**44**	44	44	44	44	44	44	**44**	44	44	44	44	44	44	44	44	44	44	**44**	44	44	44	44	44	44	44	44	44	44	44
45	45	**45**	45	**45**	45	45	45	**45**	45	45	45	45	45	**45**	45	45	45	45	45	45	45	45	45	45	45	45	45	45	45	45	45	45
46	**46**	46	46	46	46	46	46	46	46	46	46	46	46	46	46	46	46	46	46	46	46	**46**	46	46	46	46	46	46	46	46	46	46
47	47	47	47	47	47	47	47	47	47	47	47	47	47	47	47	47	47	47	47	47	47	47	47	47	47	47	47	47	47	47	47	47
48	**48**	**48**	**48**	48	**48**	48	**48**	48	48	48	**48**	48	48	48	**48**	48	48	48	48	48	48	48	**48**	48	48	48	48	48	48	48	48	48
49	49	49	49	49	49	**49**	49	49	49	49	49	49	49	49	49	49	49	49	49	49	49	49	49	49	49	49	49	49	49	49	49	49
50	**50**	50	50	**50**	50	50	50	50	**50**	50	50	50	50	50	50	50	50	50	50	50	50	50	50	**50**	50	50	50	50	50	50	50	50
51	51	**51**	51	51	51	51	51	51	51	51	51	51	51	51	51	**51**	51	51	51	51	51	51	51	51	51	51	51	51	51	51	51	51
52	**52**	52	**52**	52	52	52	52	52	52	52	52	**52**	52	52	52	52	52	52	52	52	52	52	52	52	**52**	52	52	52	52	52	52	52

Fig. 4.13 Multiple ratio on the frieze

(1;2) ... (1;10), (10;1) ... (10;10) prepared in advance. (2;3) goes in the medium box that we label inferiority ratios and (3;2) in superiority ratios.

To sort the medium boxes, several criteria are proposed, but which remain in the additive register: larger by one unit, by 2 ... This is a transition from qualitative to quantitative, but it is not the choice of the Pythagoreans who wish to measure Number1 in terms of Number2, that is to say, to use Number2 as a new unit or standard. In the case of inferiority ratios, Number1 is only a part of Number2. Pupils easily talk about fractions in this case. In the case of superiority, it is much less natural, and indeed contrary to common sense. The Greeks focused on this type of ratio to name them, so it is naturally that we now only work on superiority ratios.

Second Session: Meaning of Multiples/Divisors

This session begins with a reminder about Pythagorean figurative numbers, and a new category of figurative numbers is introduced, rectangular numbers. A number is rectangular if it can be represented by tokens in a rectangle, including a square. It is important to emphasise this point for what follows, because in everyday language, we call a rectangle what is not a square as defined by Euclid in the *Elements*, Book I, definition 22: "Of quadrilateral figures, a square is that which is both equilateral and right-angled; an oblong that which is right-angled but not equilateral." Oblong is an other word for "rectangle" (Euclid, 1956, p. 188). Not all numbers are rectangular, which is easily seen by examining the integers from 2 to 10.

There are no natural series of rectangles in the literature, because finding an order of enumeration of rectangles requires working on both length and width. As soon as the measure of one side is fixed, we can order according to the size of the other side. This is exactly what we are going to do with the series of multiples.

Composite Number/Prime Number

But before that, it is good to give a new property to numbers: being prime or being composed from the property of rectangularity. We say that a number is *prime* if it is not rectangular, *composite* if it is rectangular. It must be emphasised that *being prime or being composite* is a property of a number in itself. We can explain the term *prime* : for higher arithmetic, the one studied by the Greeks, prime numbers are what atoms are for chemistry, they are the basic building blocks of the system. The fundamental theorem of arithmetic says that every integer (in the Greek sense, that is, at least equal to two) is a product of prime numbers, not necessarily all distinct. Thus, the composite number, in a figurative representation by a rectangle, naturally relates a number to two other numbers, those represented by its sides. Since relations of three are often more difficult than those of two, we will be content with the ratio of a composite number to one of its sides. We have chosen the number corresponding to the horizontal side, which we will henceforth call *base*.

Multiple Ratios

We give the following definition: the ratio (Number1;Number2) is a *multiple* ratio if there is a figurative representation of Number1 that is rectangular with base Number2. In this ratio, Number1 is a *multiple* of Number2, and Number2 is a *divisor* of Number1.

The Pythagoreans do not consider a number to be its own multiple. This is still in line with common sense. The current mathematical meaning of "multiple" includes the equality ratio for technical reasons related to the properties of number sets, in particular the interest of neutral elements. It also includes 0, which does not exist for us, but which is a real multiplicative monster, because as an absorbing element, it is a real numerical cannibal.

Similarly, in the common sense, a divisor of a number is neither one nor the number itself, contrary to the mathematical sense which includes both. Emphasising the particular role of 1 among the Ancients allows for a better understanding of its role as a neutral element that allows it to appear and disappear at will in simplifications and fractional calculations, the resolution of equations... Not stating this explicitly can create real blockages for some pupils. This is what future primary school teachers regularly tell us. We can then distinguish the usual sense by talking about *proper multiple*, a multiple ratio always being a proper multiple ratio, since the case of equality constitutes a class already constituted.

On the other hand, in ratios, 1 is present, always in its role as leader, and for the necessity of calculations.

After checking on (10;5) and (10;4) that the definition has been understood, we proceed to sort between multiple ratios and other ratios in the box of superiority ratios. Then we lead the pupils to classify the multiple ratios by grouping together the rectangles that are in a relation of enlargement/reduction like (8;4), (4;2) and (2;1), which are *double* ratios, naturally associated with the number 2, or (9;3), (6;2) (3;1), which are *triple* ratios, naturally associated with the number 3 .

Links Between Composite/Prime Numbers and Multiple Ratios
Unlike multiplication tables, which are generally tables of the first ten multiples of the first ten non-zero numbers, our multiple tables contain the sequence of all numbers up to a given limit. Different supports are possible: tables on 10 columns, the frieze (Fig. 4.13) or an Ulam spiral (Fig. 4.12). In the project carried out, the chosen support was a 10 × 10 table, for the numbers from 1 to 99, filled in starting with the first line filled from left to right (Latin reading direction) then moving on to the next line. This support distinguishes the divisors of the base 10, which can be an advantage or a disadvantage (Fig. 4.10).

Construction of Multiple Tables
Tables 10 × 10 were distributed to each group. The instruction is to construct for each number up to 99 the table of its multiples by circling the number and colouring its multiples on a blank table (see on the site). Everyone drew the multiples of 2, 3, 4 and 5. The table of multiples of 2 revealed vertical bands, which some have reproduced for the multiples of 3 (Fig. 4.10). Attempts to rectangularise the figured numbers 10 and 13 allowed a change of strategy and the discovery of the correct method: jumping 3 by 3, then 4 by 4... Then each group was given a dozen tables to make. The last table made was that of 51, because everyone understood that afterwards, the multiples of the number "were no longer in the table and that lines had to be added to the tables".

Reading of Multiple Tables
The pupils are then invited to study the tables and say whether these supports can be used to answer the following questions:

- How to verify that a number is prime?
- How to find the divisors of a number?
- How to verify if (Number1;Number2) is in multiple relation or not?
- How to find the name of the multiple ratio, i.e. how many times Number2 is in Number1?
- How to verify if Number1 and Number2 have a common divisor?

It is always interesting to compare the methods induced by the different media. We can state properties such as: (i) *a number is prime if it does not appear coloured in any table of multiples of a number smaller than it*; (ii) *ratio is multiple if* Number1 *is coloured in the table of multiples of* Number2; (iii) *To find the divisors of a number, you have to look in each table of multiples of numbers smaller than it to see if it appears coloured.*

The properties would have been stated identically with the Ulam spirals, but differently using the frieze (Fig. 4.13). Indeed, while the tables and spirals allow a linear reading of the numbers, the frieze offers a two-dimensional reading: each column gives the multiples of the first bold number (the first isolated column is the list of multiples of one), each line the list of divisors of the number that is on the line and on the first column, which is also the last bold number of the line (reading from left to right).

To find the designation of the multiple ratio, some pupils suggest doing the division, while others point out that it is enough to count its rank in the coloured numbers starting from 1 for Number1. This remark allows us to explain that in mathematics, it often happens that for small cases, or for certain special cases, we find the result more easily directly or with methods seen when we were little, and that these non-expert methods, allow to find the expert methods (of "adults"). These expert methods are essential for all the numbers. For example, from the age of 11, knowing the calculation tables and a method for each operation allows to go faster than to build tables of multiples, if only to study the ratio (1, 273, 358; 2018), even by only writing the multiples of 2018, you have to write 630 to answer.

Between January and March: Classification of Ratios and Fractions

Materials

All the media used previously are available to pupils.

- Four boxes of the same size as those of the multiples.
- A summary sheet of the different classes of ratio (Fig. 4.14), either one sheet per group, or one poster for the class.
- A table for the teacher (Fig. 4.16) in order to quickly have available pairs of prime numbers between them, and the same in black and white to cut out the ratios, other irreducible ratios if needed.
- Flexible unit-graduated rulers with or without guide-donkey (Fig. 4.6).
- Questionnaire sheets (Fig. 4.15).

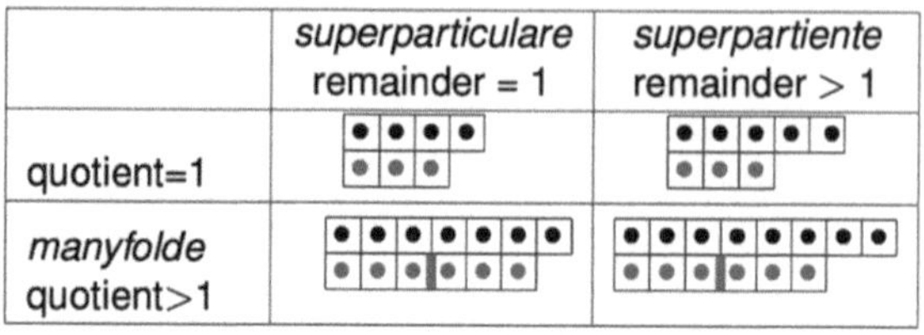

Fig. 4.14 Classification by type of superiority ratios of numbers

Team's first names :

(NUMBER1= 46 ; NUMBER2=7)

NUMBER1=q times NUMBER2 plus r $46 = 6 \times 7 + 4$ reminder : for $r \neq 0$, IF r=1 THEN particulare ELSE partiente	its class : ☐ Multiple ☐ Superparticular ☐ Superpartiente ☐ Manyfolde superparticulare ■ Manyfolde superpartiente

Its name is : *sextupla superquadrupartiens septimas*

Fig. 4.15 Class sheet of ratio (data: two prime numbers between them)

First Session: Framing a Number Between Two Successive Multiples
Let a and b two numbers such that b is smaller than a. It is a question of framing a between two successive multiples (including b itself) of b. For example, 33 and 7: 33 is between 4 times 7 and 5 times 7, which can be read easily on the multiple table of 7 if you do not know the table of 7. Excess or defect of a is always smaller than b. It can be noted that in everyday life, it often happens that it is the higher multiple that one must take, such as when buying eggs in a box. Mathematicians have chosen to favour the lower multiple. This session is an important link in the sequence.

In the project, it was decided not to present it within the framework of the activity, but as a standard mathematics course in order to test the effectiveness of the previous sessions. The multiple table supports were used for small numbers by the pupils at their request. This session was very easy for 80% of the pupils including the youngest ones (9–10 years old), which was a very good result for the teachers. The pupils then easily specified in each case studied, what remained in excess compared to the lower multiple and by default to the higher multiple.

Second Session: Number Ratios and Fractions
In this session, we return to our Pythagoreans. We saw during the last session of the November-January period that for every multiple ratio, we can associate an integer which is the ratio of Number1 to Number2, which we call in mathematics the quotient of Number1 by Number2. This integer is easily identifiable on a graduated

ruler in units. The aim of this session is to show that we can also associate a point on the graduated ruler in units with any number ratio.

Locate by folding: the ratios Number1=2 times Number2 +1
We ask the pupils to recall the framing of 3 and 9 in relation to 2 and its multiples, then where they are located in the multiple tables. They are located in the middle. We then ask them where they would locate the ratios (3;2) and (9;2) on the graduated ruler in units. To preserve this situation, (3;2) is located in the middle of 1 and 2, (9;2) is located in the middle of 4 and 5. They therefore correspond to numbers—called broken by mathematicians until the seventeenth century—1 1/2 and 4 1/2 which we say *one and a half and four and a half* implied *unit*. We find these notations in some cooking recipes, and English pupils naturally represent fractions in this way. We write mathematically: $\frac{3}{2} = 1 + \frac{1}{2}$ and $\frac{9}{2} = 4 + \frac{1}{2}$, which exactly describes the placement of the ratios on the ruler and is obtained from the two equalities seen in the previous session: $3 = 2 \times 1 + 1$ and $9 = 2 \times 4 + 1$.

Pupils are then asked how one can "mark" the middle between two successive graduations. One can use a compass and mediator, but the simplest method remains folding. The same process applies to the ratios (5;4) and (17;4) which require two successive folds once the framing is done.

In mathematics, we write $5 = 4 \times 1 + 1$ and $\frac{5}{4} = 1 + \frac{1}{4}$, $17 = 4 \times 4 + 1$ and $\frac{17}{4} = 4 + \frac{1}{4}$.

We then ask the same thing for (6;4) and (18;4), which are—unlike the others—reducible ratios. 6 is located in the middle of 4 and 8; 18 in the middle of 16 and 20. Therefore, (6;4) must be located in the middle of the positions of 1 and 2, so the ratios (6;4) and (3;2) represent the same position and therefore the same broken number. (18;4) in the middle of the positions of 4 and 5 so the ratios (18;4) and (9;2) represent the same position therefore the same broken number.

In mathematics, we write $6 = 4 \times 1 + 2$ and $\frac{6}{4} = 1 + \frac{1}{2}$, $18 = 4 \times 4 + 2$ and $\frac{18}{4} = 4 + \frac{1}{2}$.

In these examples, we were able to show the pupils that, as with multiple ratios, there are several ratios that represent the same thing. The Pythagoreans gave the same name to ratios that represent the same thing. For example, they named the ratios of 3 to 2 and 6 to 4 sesquialter (*cf.* section "Classification of Non-multiple Ratios", p. 97), and the ratios of 9 to 4 and 18 to 8 *sesquiquadriple* (Table 4.5).

We then agree with the pupils that we will now only work on irreducible ratios.

Locate any Irreducible Ratio
We propose to the pupils the guide-donkey, verifying with them that to divide a segment in two, the guide-donkey works well too. Then we propose to divide a segment into 7 equal parts with folding and guide-donkey (Figs. 4.6 and 4.7). It's simpler with the guide-donkey. Then, we point out that as soon as the number of parts is greater than 2, there are remainders that will be larger than 1, which will require taking several parts, as for the ratio (31;7). As $31 = 7 \times 4 + 3$, the ratio (31;7) or the fraction 31/7 is located between 4 and 5, exactly 3/7 away from 4 and we write $\frac{31}{7} = 4 + \frac{3}{7}$.

Fig. 4.16 Classification by type of number ratios from the treatise of Chuquet

6	3	3	2	11	3
6	2	4	3	17	5
8	2	5	4	15	4
10	2	6	5	18	5
5	2	7	6	19	5
7	3	5	3	25	7
9	4	7	5	14	3
7	2	7	4	23	5
10	3	8	5	24	5
13	4	9	5	32	7
9	2	11	7	17	3
13	3	8	3	27	5
17	4	12	5	23	4
11	2	14	5	23	5
16	3	18	7	28	5
21	4	33	5	29	5
13	2	46	7	39	7
19	3	27	4	20	3
25	4	34	5	32	5

As soon as a group has correctly located (31;7), it receives another irreducible ratio, taken from Chuquet's ratio table (Fig. 4.16).

Third Session: Complete Classification of Superiority Ratios

We start by recalling that the Ancients, up until the seventeenth century, treated 1 differently from other numbers, which means that equality ratios are not multiple ratios. What remains are the non-multiple superiority ratios to be dealt with. The pupils have well understood that these ratios correspond to a Number1 framed (strictly) either by Number2 and its double, or by two of its successive multiples. Then, on the excess (or the remainder), one must distinguish whether it is worth 1 or more. How many different boxes will we have? There will be four boxes, described in Fig. 4.14.

In possession of Fig. 4.16, which lists the superiority ratios of irreducible numbers expressed by Chuquet (1880), the teacher fills in for each group the line (Number1;Number2) of a card represented in Fig. 4.15.

In the experimental classes, most pupils quickly mastered the management of the two variables for the classification, but some were bothered by the fact that a non-multiple ratio was in a class also carrying the term multiple which they found misleading. Others pointed out that with something behind—an adjective that determines it—it was like saying "almost multiple". The answer has a flavor of Marie Poppins' *supercalifragilisticexpialidocious*.

As soon as a group came back with its card properly filled out, it could submit the ratio and put it inside its box.

Ways to Go Further

We outline here two paths: the first concerns the naming of ratios, the second concerns instructions related to classification.

Playing with Words

An additional session can be added for pupils over 11 years old, to deal with reducible ratios.

With the help of cards giving one element of the ratio and the name like:

> *My name is (super-particular) sesquialter, I have lost my first member, help me to rebuild myself, I am (– ;14)*

We can also work on proportional ratios (or fractions). Examples of cards are proposed on the site.

We can also form a word from a reducible ratio, for example (9;6). As $9 = 6 + 3$ we ask if "supertriparticularsixth is the name of a ratio?", the winner is the first to answer "no, it is the ratio sesquialter".

In connection with the French teacher [resp. the English one], we can work on the original text of Chuquet [resp. Recorde (1557)] or a translation in present-day language.

We can encourage mental calculation on fairly small numbers during flash questions that ask both the type of ratio and its expression in the form of an integer increased if necessary by an irreducible fraction smaller than the unit.

Elements of Coding and Programming

The classification by genre is based on cascading conditional instructions. Writing the algorithm for coding a ratio into its name and vice versa can be done from the first lessons of algorithmic, both in a mathematics course from middle school and in a computer science course in high school. Reflecting on the implementation, in particular the fact that it is necessary to give the machine, in addition to the construction rules, the vocabulary of Table 4.5 without the ellipsis is part of the computer science course. We can always give the machine a ratio whose name it will not know, while a person will be able to do so if she has understood the rule.

Acknowledgments This chapter was written by Sylviane R. Schwer with the help of the primary school teachers Yan Guede, Deborah Fau, Catherine Levy, Grégoire Charles, Tiffany Gérin, and Saranya Zearamane from Ormeteau School (Aulnay-sous-Bois, France), and Fabienne Romian, a mathematics teacher at Gérard-Philippe Middle School (Aulnay-sous-Bois). They all followed the experiments during the 2016–2017 school year.

Appendix: Concepts and Skills Addressed in This Chapter

Concepts Covered

- Calculating with whole numbers
  - Multiples and divisors.
  - Enclose a number between two consecutive multiples.
  - Find a quotient, a remainder.
  - Solve proportionality problems (reducible, irreducible ratios).
- Representing fractions
  - Understand the concept of fraction (simple).
  - Various designations of fractions.
  - Enclose a fraction between two consecutive whole numbers.
  - Establish equalities between simple fractions.
- Organisation and data management
  - Produce tables organising numerical data.
  - Produce geometric figures representing numbers.
- Reasoning and algorithmics
  - Data representation.
  - Case reasoning and conditional instructions.
- Intersections between teachings
  - Classification.
  - Rules for constructing a specific vocabulary to designate number ratios.

Skills Developed

- **Search and research skills**: Observe various representation supports to conjecture properties.
- **Reasoning**: Case reasoning based on different constraints. Understand the meaning of a Euclidean division and exploit the different elements of a Euclidean division to name a ratio.
- **Computational skills**: Perform Euclidean divisions.
- **Mathematical communication skills**: Understand the formation of series of terms related to numbers.
- **Mathematical representation ability**: Use different types of support to express the same property in different ways.
- **Modelling**: Choose the right representation depending on the problem. Distinguish between a property of an object (number, pair of numbers) and a relationship between two objects.

References

Boethius. (1983). *De institutione arithmeticæ*. Rodopi.

Brissiaud, R. (1998). Les fractions et les décimaux au CM1, une nouvelle approche. In *Actes du XXVe colloque des formateurs et professeurs de mathématiques chargés de la formation des maîtres* (pp. 147–171). IREM de Brest.

Caiazzo, I. (2021). Medieval Commentaries on Boethius's *De arithmetica* : A provisional Handlist. *Bulletin of Medieval Philosophy*, *62*, 3–13.

Chuquet, N. (1484/1880). *Le Triparty en la science des nombres*. Imprimerie des sciences mathématiques et physiques. http://gallica.bnf.fr/ark:/12148/bpt6k62599266

Euclid. (1956). *The thirteen books of Euclid's elements*. Dover Publications. Volume I, Introduction and Books I-II (T. Heath trans.), 2nd edition.

Fux, J., & Miller Schulte, S. (1725/2017). Gradus ad Parnassum. A Reader's edition and commentary by Sara Miller Schulte. Master's thesis, Western Michigan University. https://scholar-works.wmich.edu/masters_theses/1993

Getenet, S., & Callingham, R. (2017). Teaching fractions for understanding: Addressing interrelated concepts. In A. Downton, S. Livy, & J. Hall (Eds.), *40 years on: We Are Still Learning! Proceedings of the 40th Annual Conference of the Mathematics Education* (pp. 277–284). MERGA.

Gibson, M. (1982). Boethius in the Carolingian Schools. *Transactions of the Royal Historical Society*, *32*, 43–56.

Hilbert, D. (1897/1998). *The theory of algebraic number fields*. Springer.

Høyrup, J. (2014). Mathematics Education in the European Middle Ages. In A. Karp & G. Schubring (Eds.), *Handbook on the history of mathematics education* (pp. 109–124). Springer.

Jean Abraham, known as Launay. (1665). *L'Arithmétique, arpentage universel, toise des bastimens, géométrie inaccessible, géométrie par la règle et le compas, avec la fabrique et usage des quadrans sollaires*. L'Oyselet.

Laisant, C.-A. (1907). *La Mathématique. Philosophie, Enseignement*. Gauthier-Villars.

Poincaré, H. (1913). *The foundations of science*. The Science Press. https://www.gutenberg.org/files/39713/39713-h/39713-h.htm

Recorde, R. (1540/1582). *The ground of arts: Teaching the perfect work and practice of Arithmetic, both in whole Numbers and Fractions, after a more easier and exact form than in former time has been set forth*. I. Harrison & H. Bynneman. https://quod.lib.umich.edu/e/eebo2/A10530.0001.001?view=toc

Recorde, R. (1557). *The Whetstone of Witte, whiche is the seconde parte of Arithmetike*. Ihon Kyngstone. https://www.loc.gov/item/04006305/

Sylviane R. Schwer is a discrete mathematics researcher at the Laboratory of Informatics of Paris Nord (LIPN, UMR 7030) and a mathematics teacher at Sorbonne University Paris Nord, very involved in teacher training. Her main research focuses on the representation in language of the numeration system and calendar time. She is also interested in the history of combinatorics and probabilities of the interwar period 1870–1914, especially the work of Henri Delannoy that she helped to make known.

Part II
Magnitudes and Measures

Chapter 5
Doubling the Square with Plato

Renaud Chorlay

Abstract This chapter bears on a famous problems discussed in Plato's *Meno*: for a given square, construct a new square with twice the area. We designed and tried out a series of three work sessions for pupils of grades 5 and 6 (aged 10–11) which give pupils the opportunity to engage with three key and challenging mathematical concepts: area, proportionality, and the characterization of common quadrilaterals. A key design choice is that pupils were not only asked investigate the problem from the *Meno*, but also to read large extracts from this original text, thereby reflecting on the mathematical problem while endeavouring to make sense of the text. Consequently, the design choices rest on inputs from two research fields: mathematics education and didactics of reading comprehension.

Keywords Plato · Area · Reading comprehension · Proportional thinking

Introduction

We propose in this chapter a sequence of three sessions, designed and tested in years 5 and 6, based on one of the most famous problems in the history of mathematics. In the dialogue titled *Meno* (Plato, 1924), Plato (*circa* 427–348 BC) writes an exchange between three characters, the philosopher Socrates, a noble Athenian named Meno, and a slave in Meno's service. Socrates (*cf.* Fig. 5.1) starts with a square (a) and asks the slave to draw a square with double the area. The slave finds the question easy, and proposes to double the side of the square (state (b)). Socrates shows him that this solution is wrong: figure (b) is indeed a square, but its area is four times that of the square (a), as shown in figure (c). Socrates then connects four points of figure (c), and asserts that the square that appears is—indeed—double in area of the

---

R. Chorlay (✉)
IREMS de Paris, Université Paris Cité, Paris, France
e-mail: renaud.chorlay@inspe-paris.fr

M. Moyon et al. (eds.), *Teaching Mathematics Through Historically-Based Activities*, History of Mathematics Education,
https://doi.org/10.1007/978-3-032-16990-7_5

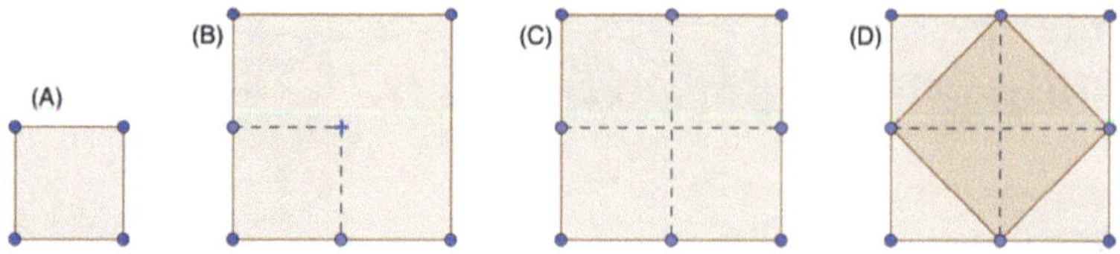

Fig. 5.1 The successive figures drawn in the *Meno*

initial square. He justifies it by counting in both figures (a) and (d) the isosceles right triangles that are the half-squares: two in the square (a), four in the square (leaning) in the centre of (d).

We see that the problem focuses on fundamental concepts and themes of work at the primary-secondary transition: use of precise vocabulary to describe fairly complex geometric figures; characterisation of usual quadrilaterals (in particular, to verify that the "tilted" quadrilateral is indeed a square); comparison of areas, either without measurements—by cutting and reassembling—or by measuring, sometimes with conventional units (here the foot) sometimes with units chosen according to the situation (here the half-square (a)). Moreover, the slave's mistake allows us to work on two epistemological obstacles: difficulty in distinguishing, for the same plane figure, two different quantities (the length of the outline, the area of the surface); non-proportionality of lengths and areas in enlargement/reduction situations.

Our approach in the construction of the sequence was to have the pupils work not only on the *problem* of the *Meno*—doubling (in area) a given square—but on the text of the *Meno* (Plato, 1924). Our motivation was twofold.

Firstly, this seemed to us to encourage the entry into argumentative work in geometry, in a context where the teaching of mathematical proof is not yet targeted: on the one hand, the text is of an *argumentative* nature; it presents exchanges of arguments—between Socrates and the slave—about figures and magnitudes. On the other hand, the work entrusted to the pupils will consist not only in seeking the answer to the posed problem, but also in appropriating and evaluating the arguments exchanged in the text.

Secondly, we wanted to acknowledge the fact that this text provides the opportunity for interdisciplinary work in mathematics and language mastery (*reading comprehension*). We detail this aspect further on.

Historical Context

The few historical references that we give here are intended for the teacher. They do not play an explicit role in the sequence proposed to the pupils.

As for the state of mathematics in the Greek world at the time of Plato (*circa* 427–348 BC), we invite the reader to consult Burton (2011) or Netz (2003). The oldest mathematical texts that have come down to us are later than this period: Euclid's *Elements* and Autolycus of Pitane's *The Moving Sphere* were written around 300 BC. The *Elements* are, however, already a synthesis work; they attest to the existence of a well-established research tradition, focusing on abstract questions related to geometric figures and magnitudes (plane and solid), integers, proportions, links between magnitudes and numbers. Of the mathematical works of Plato's predecessors or contemporaries, we only have either indirect traces—for example for those of Hippocrates of Chios—or more of the fable than of history—for example for the measurement of pyramids by Thales. However, we know that the questions of area comparisons are studied. Thus, Hippocrates works on the areas of lunules (surfaces delimited by two arcs of a circle), seeking to solve the problem of *quadrature*, that is, the construction of a square of the same area.

At the time of Plato, these mathematical questions are therefore discussed in intellectual circles, although probably marginally compared to more noble themes (philosophy, ethics, politics), and within the restricted frameworks of circles and schools linked to thought leaders. Several Platonic dialogues give a role to mathematics. Plato emphasises their importance in the training of the elites of the *Republic*. The cosmological reflections of the *Timaeus* rely on several mathematical elements: association of the four elements—fire, earth, water, air—with regular polyhedra; neo-Pythagorean valorisation of certain number ratios.

Mathematics, however, is not the central theme of the *Meno*, any more than of the *Republic* or the *Timaeus*. The dialogue between the philosopher Socrates and the nobleman Meno initially focuses on the search for a definition of "virtue" or "excellence". After several attempts at definitions have been proposed by Meno and refuted by Socrates, the dialogue takes a more epistemological turn. It examines what it is to know—what virtue is or anything else—and how it is possible to acquire knowledge. The episode that will serve as the basis for our mathematical work is a digression proposed by Socrates: He relies on the journey of the slave confronted with a mathematical problem to illustrate his theory of access to knowledge: distinction between ignorance (lack of knowledge) and illusion (belief in a "knowledge" that is erroneous); the need to become aware of one's ignorance and illusions in order to, in a second step, access the truth; the role of dialogue *maieutic* not to transmit knowledge from the wise to the ignorant, but to bring to the consciousness of the one who seemed ignorant—the image is that of childbirth—the truths and knowledge that were in him, although hidden. This access to truths is described as a *reminiscence*, remembrance of truths contemplated by the eternal soul before birth. In a sense, according to this theory, we never learn anything; we can remember, if we are prepared and guided. We see that the pedagogical model illustrated by the progressive resolution of a mathematical problem is rooted in a general epistemological questioning, to which a response is provided involving cosmological elements.

Scientific and Pedagogical Aspects

As far as mathematics is concerned, the notions and knowledge at stake in the problem of the *Meno* are easy to identify: use of precise vocabulary to describe fairly complex geometric figures; characterisation of usual quadrilaterals (in particular, to verify that the "tilted" quadrilateral is indeed a square); comparison of areas, either without measurements—by cutting and reassembly—or by measuring, sometimes with conventional units (here the foot) sometimes with units chosen according to the situation (here the half-square (a)). Beyond the list of notions, let's specify some choices that guided us in the construction of the sequence.

The general perspective is socio-constructivist, in the sense that the work on the notions goes through the interaction of the pupils with a mathematical environment presenting several characteristics: it is familiar enough to the pupils for them to be able to appropriate the stakes and enter into the research; the complexity of the task and the non-obvious nature of the solution give the environment a certain resistance; the pupils have elements to evaluate—at least in part—the validity or invalidity of their actions or proposals. The sequence we propose is however not designed as a situation aiming at the discovery or the first encounter with the notions at stake, at least as far as the instrumented characterisation of the square, and the comparison of areas with or without measurements are concerned. On the contrary, the situation regularly puts the pupil in the position of an expert who must rely on the knowledge of these notions not only to try to solve a geometric problem, but also to *reformulate*—or even criticise—formulations sometimes far from ours, as well as to *evaluate* the quality of arguments proposed in the text.

The sequence provides an opportunity to confront pupils with epistemological obstacles well identified in the didactic literature: difficulty in distinguishing the two quantities associated with the same surface (length of the contour, area); prevalence of the linear model, yet in default in situations of enlargement/reduction (here: doubling the side of a square does not double the area). Unlike the knowledge on quadrilaterals and area comparisons—on which a certain familiarity from pupils is expected—the sequence cannot assume any expertise from the pupil on these delicate points. Always from a socio-constructivist perspective, the construction of the sequence aims not to evade these difficulties, but on the contrary to bring the pupils to confront them in order to become aware of them. Our objectives here are modest: we would like that at the end of the sequence the pupils can say that by doubling the side of a square one does not double its area, justifying it by the exhibition of a counter-example. We will now give elements on what happened during the two experiments.

In the problem of the *Meno* one can see one of the oldest formulations—on a particular case—of the so-called Pythagorean property which, in the Greco-Latin tradition, links the areas of the squares built on the three sides of a right-angled triangle (Euclid's *Elements*, Book I Prop. 47 (Euclid, 1956, pp. 349–368)). The contemporary reader may have learned it as a *numerical* relationship concerning the measurements of the lengths of these sides. This reader may be surprised not to see in our sequence of work numerical work on the side of the square sought. Mathematically, the sought square (that of double area) and the given square have sides in the ratio of $\sqrt{2}$ to 1. Neither the square root, nor the Pythagorean theorem—which can be used to determine the length of the diagonal of a square in relation to that of its side—are retained as objectives for our teaching sequence.

The main characteristic of the sequence we propose is that it involves work on the text of the *Meno*, and not just on the mathematical *problem* present in this classic. Therefore, the characteristics of this text had to be taken into account when designing the session. It is a long text (6 pages); it is an argumentative text. It is a long digression—with mathematical content—within a longer text of a philosophical nature. The intertwining of the two scales—the digression/the complete dialogue—and the two themes—mathematical problem/philosophical argumentation—structures the extract we study: the dialogue between Socrates and the slave (on the mathematical problem) is regularly interrupted by asides between Socrates and Meno, where the latter debate—in a sibylline way—what the slave's journey in the mathematical problem brings to their philosophical discussion. In the mathematical study, the situations are not symmetrical: the slave discovers *at the end* that his first solution is wrong, while Socrates and Meno know from the start that he is mistaken; for them, the issue is not about the areas of squares, but about what "believing", "knowing" and "learning" mean.

For these reasons, not only is the text long but it is objectively *difficult* to understand, and the mathematical difficulties are not the only ones at stake. Understanding the text requires not only mathematical work on figures and magnitudes, but also regular work of explanation, reformulation, formulation of hypotheses about the knowledge and intentions of each of the three characters. In addition, many passages remain obscure at first reading, because the key to understanding them comes later in the text. The overall meaning of the text—and the precise meaning of certain passages—is therefore gradually revealed, through successive returns to the path taken, through a revision of reading hypotheses. This has a cognitive and affective cost: one must accept not to all understanding straight away; one must accept that at different times during the work, different parts of the text are subject to different degrees or qualities of understanding.

Choosing to have pupils work on the text of the *Meno* is choosing to confront, as a teacher, this difficulty of the text. This is not a whim on our part. We are encouraged to do this by the work on the didactics of text comprehension.

We sought to design the sequence based on the analysis of the cognitive and didactic challenges of text comprehension as described by the team of authors of the manuals *Lector & Lectrix* (Cèbe & Goigoux, 2009). They remind us that "understanding a text" encompasses several families of skills: decoding the written code (the usual meaning of "knowing how to read"), implementing linguistic and textual skills (syntax, lexicon, punctuation, connectors, etc.), referential skills (here the text is—largely—*about* mathematical objects), and finally, strategic skills (regulation, control and evaluation, by the pupil, of his reading activity). Cèbe and Goigoux emphasise:

> If he wants to understand a text, the reader must mobilise all these skills simultaneously to carry out two main types of processing: local processing—which allows him to access the meaning of groups of words and sentences—and more global processing that leads him to construct a coherent mental representation of the whole. [...] This latter process, called semantic integration, is cyclical: each set of new information leads the reader to reorganise the representation he is building step by step [...] This assumes that he is flexible enough to accept that his initial representations are provisional and therefore revisable. (Cèbe & Goigoux, 2009, p. 7)

The characteristics of the *Meno* led us to construct the sequence in such a way as to accompany the pupil—and the whole class—in this process of semantic integration. We take from *Lector & Lectrix* pedagogical guidelines that allow this scaffolding; let us mention in particular (adapted from Cèbe and Goigoux (2009, pp. 17–20)):

- Make pupils active and able to regulate their reading. [...] We invite them to evaluate their degree of understanding ("I am sure", "almost sure", "not sure at all".)
- Encourage filling in the blanks of the text: one must cooperate with the text to go a little beyond what it explicitly says, while respecting the "rights of the text".
- Lead to questioning the thoughts of the characters: [...] their goals (for the future) and their reasons for acting (in reference to the past); their feelings or emotions; their knowledge and reasoning.
- Encourage recall and reformulation to learn to memorise: we multiply the tasks of paraphrasing and reformulation.
- Make vocabulary a permanent objective: explanation by the teacher before or during reading; but also, awareness of the possibility for the reader to make hypotheses about the meaning of an unknown word.

The combination of difficulties related to mathematics and the text comprehension can be daunting. It is up to us to prove that we have been ambitious but not reckless. The sequence was implemented in regular classes of levels 5 and 6, in the spring of 2017. It takes three sessions; plus—if desired—a small preliminary study of documents, which can be conducted in conjunction with the history course. We will provide in the following elements on the actual course in the two classes. These elements are based on the data collected in classes: collection of pupil worksheets, transcription of audio recordings of the sessions.

That being said, the version of the sequence that we present here assumes a maximalist, even experimental side. Everyone will be able to adapt it according to their desires. To the teacher wishing to have his pupils work on the *problem* of the *Meno* without having them work on the text itself, this report will provide elements on the procedures, responses and difficulties encountered by pupils on the occasion of the study of this problem.

Implementation in the Classroom

It seemed to us that the text could quite naturally be divided into three parts, each leading to a work session of about an hour.

- 1st session: presentation of the context. Discovery of the text.
- 2nd session: discovery of the duplication problem. Comparison of the pupils' responses and that of the slave. Invalidating the response by doubling the side.
- 3rd session: discovery of Socrates' response. Validation or invalidation by the pupils of this response. Review of the general construction of the text and its philosophical meaning.

First Session

A small introduction seems necessary to set up certain contextual elements and avoid certain misunderstandings—for example on the question of slavery. The following points are discussed with the pupils: we are going to study a text written 25 centuries ago, in Antiquity, in Ancient Greek, by an author named Plato (*circa* 427–348 BC), active in Athens. Plato is a famous author for his philosophical texts—that is to say, reflective. In Plato's time, the political system in Athens was democracy. The population was divided into several categories: citizens—men, Athenians, enjoying civil rights—, non-citizen free people—women, foreigners in Athens—, slaves. Slaves were most often either children of slaves or prisoners of war; they worked in houses, workshops or fields.

From a time perspective, it is preferable not to exceed 10 minutes at most for setting the context, otherwise the integrity of the session may be compromised. A possible variant is to remove moment 3 (see below), or to deal with it quickly orally.

Moment 1: Discovery of a *Difficult Text*

We then distribute to each pupil the beginning of the passage on the duplication of the square (Appendix 2).

Instruction: I'm not telling you what this text is about; you will see that it is a difficult text, and that all alone you will not be able to understand *everything* at first glance; but you will see, by working together we will be able to understand this text, and it will teach us to read difficult texts. Take 5 minutes for a first reading, in silence.

Pooling together allows us to identify some very simple elements. As for the type of text, it is a dialogue, it resembles a play. There are three characters. The text seems to talk about mathematics: it talks about squares, lines, length, calculations; yet it does not resemble our mathematics textbooks. It is also noted that in the designation of the document, "1924" is probably not the date of writing; that "Lamb" is the name of the translator, an opportunity to remember that the original text is in Ancient Greek.

The text is then read again, this time aloud, by three pupils. To encourage more attentive reading and lead the pupil to evaluate their level of understanding, the following instruction is then given:

> The text remains difficult to understand. I ask you to take 3 colours and underline the text (you can write on it) using the following coding:
>
> Green: here, I understand.
> Orange: here, I understand a little but I'm not sure.
> Red: here, I do not understand.

Then, after five minutes dedicated to this work, a phase of dialogued lesson is launched by the instruction:

> The text is difficult to understand for several reasons. Who can tell me which passage is difficult to understand, and *why*?
> I repeat: I want you to try to tell me why such and such passage is a bit difficult to understand.

We had anticipated that the following correct answers and might be expected:

- First difficulty: At the beginning of the text, we do not know what Socrates and Meno are talking about. The teacher can then indicate that the text is an excerpt from a longer dialogue, and that indeed, we do not know what Socrates and Meno were talking about before. The first lines indicate that Socrates proposes to try to convince Meno of something; of what, we do not know. One of the objectives of our work, but it will only be achieved at the third session, is to understand what Socrates wants to convince Meno of.
- Second difficulty: The text is difficult because it talks about geometric figures that we do not see. The teacher can ask what in the text, allows to affirm this, an opportunity to make pupils notice the deictics ("like *this*," "*these* lines," etc.). We then ask the pupils to imagine what Socrates does while he says this: we can reasonably imagine him drawing figures, on a board, a sheet of paper, or on the dust of the ground. We agree with the pupils that, to understand the text, it would be preferable to have the missing figures in front of us.

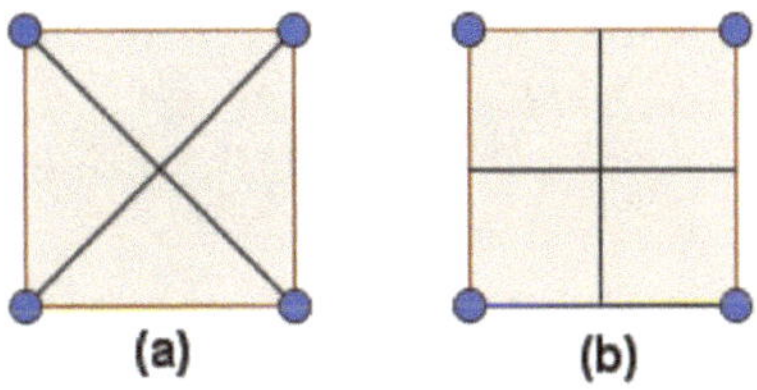

Fig. 5.2 Two figures compatible with an ambiguous text

Moment 2: Conjectures About the Missing Figures. Distinction of the Roles of Lengths and Areas

Then begins the more specifically mathematical work.

> Instruction: Now remains the problem of the geometric figure that Socrates shows to the slave. The text gives you enough information to be able to do it; it's up to you to propose a figure (or figures) that help to understand the text. You can do it (or them) on the document; you can draw it (them) freehand; if you want, you can write things on it. You have 5 minutes. You can work in pairs.

If drawing a square is not a problem—freehand work is allowed—the text raises a question. Lines that pass through the middle of the square and are equal (in length), several can be imagined (Fig. 5.2):

It is agreed that these figures are both compatible with the text. No pupil made a mistake by doing one rather than the other, or hesitating; it is the text that is ambiguous—it is an opportunity to introduce this term to the pupils, a term useful in mathematics and beyond. We can also ask the pupils to describe precisely the two figures, which provides the opportunity to use the terms: segment, side, diagonal, midpoint of a segment, segment connecting two opposite vertices, segment connecting the midpoints of opposite sides.

It turns out that the rest of the text resolves the ambiguity, in favour of figure (b). We ask the pupils—for example by coming to the board—to show on the figure where one can see "two feet long". In the classes this did not pose a problem. Many pupils know that the "foot" has been used as a unit of length; often they enjoy citing others, like the "inch". This gives greater plausibility to figure (b), without excluding figure (a).

We then write on the board the two questions:

The sentence "Then the space is twice two feet" is very important. Take a few minutes to think, and prepare in writing your answers to the two questions:

- Can you explain what Socrates is talking about?
- Do you agree with him?

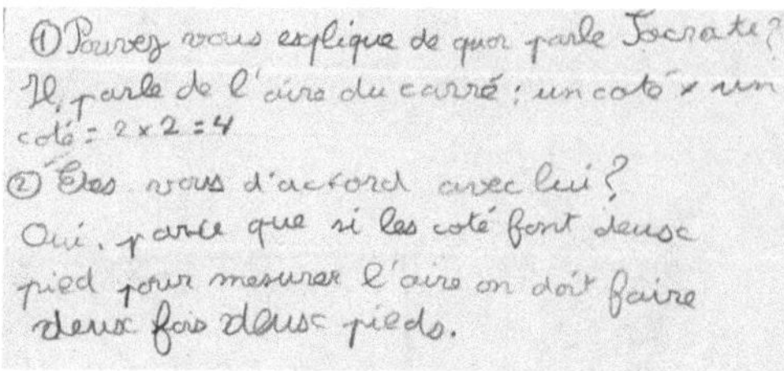

(a) Worksheet #1

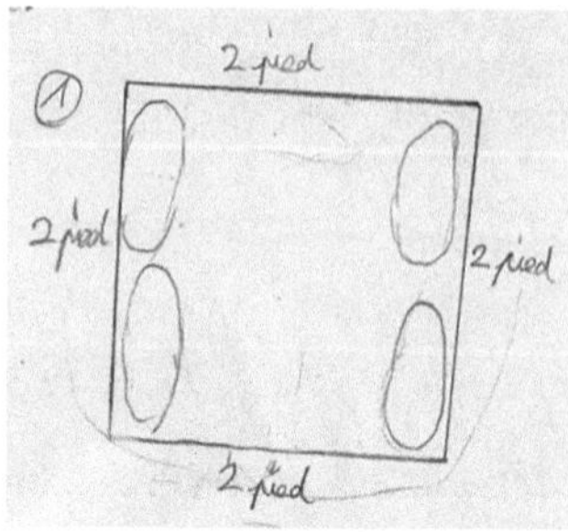

(b) Worksheet #2

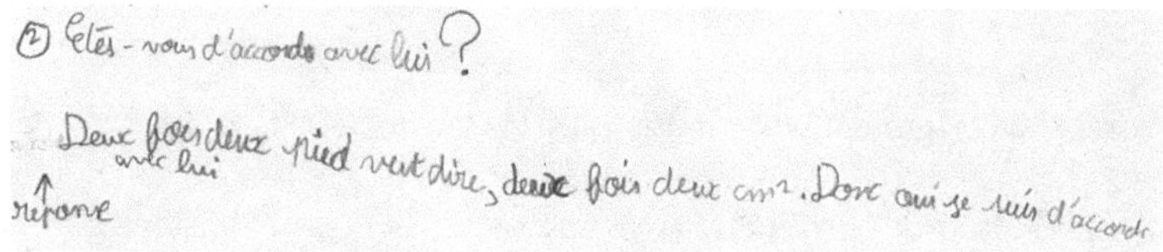

(c) Worksheet #3

Fig. 5.3 A "space of two by two feet", three interpretations. (English translation) Worksheet 1: "Can you explain what Socrates is talking about? He's talking about the area of the square: one side × one side = 2 × 2 = 4. Do you agree with him? Yes, because if the sides are two feet to measure the area we have to do two times two feet." Worksheet 3: "Do you agree with him? Answer: Two times two feet means, two times two cm^2. So yes, I agree". (**a**) worksheet n° 1. (**b**) worksheet n° 2. (**c**) worksheet n° 3

You can make sentences, or freehand figures; you can colour; everything is allowed.

The aim of these rather open questions is twofold:

- To get the pupils to see that, although the beginning of the text only talks about lines and lengths, it is the area of the square that is questioned from "how many feet would the whole be?", or more precisely the link between the lengths of the sides and the area of the square. Socrates is then right to say that if the square is 2 units of length on the side, then the square is 4 units of area; we can verify this either by counting the unit squares in the figure (which is easy on (b)), or by using the formula for the area of the square.
- To get the pupils to criticise Socrates' formulations as ambiguous, or even incorrect for the modern reader. Socrates indeed uses the same word "foot" to refer to both the unit of length and the associated unit of area.

The course of events in the classes shows that—when they reflect alone or in pairs on these questions—the pupils clearly identify the fact that we are talking about the area of the square. We reproduce extracts from three Year 6 worksheets (Fig. 5.3):

Answer 1 is representative of the majority response; many of pupils, more laconic, validate by "yes, because 2×2 = 4".

Answer 2 even shows a case of representation allowing to accommodate the ambiguity of the text: the drawn feet designate both the lengths (by their longest

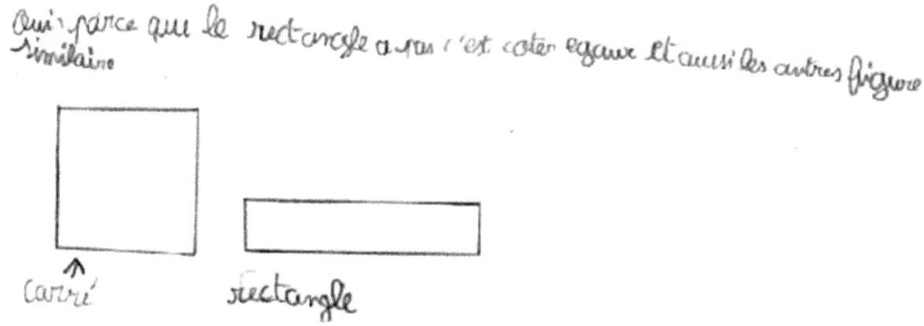

Fig. 5.4 Is a figure with four equal sides always a square? (English translation) "Yes, because the rectangle does not have equal sides and also other similar figures." (original spelling)

side) and—presumably—the area feet (by the part of the surface on which they are drawn).

Answer 3 shows a case, rare, where the pupil recalls the role of the square to designate the unit of area. We see that the collective assessment is necessary to show the problematic nature of the use by Socrates of the same term to designate the unit of length and the companion unit of area. The introduction of the term "square foot" for the unit of area does not shock the pupils; it can be proposed by some. We can also choose to distinguish the "feet of length" and the "feet of area".

Moment 3: Characterisation of the Square

Question: in Plato's text, Socrates seems to say that a figure which has four equal sides is necessarily a square. Do you agree?

In fact, the term "of the same sort" used by Plato probably refers to the mathematical notion of similarity, in which case Socrates is right. However, for the pupils, we consider the term as belonging to ordinary language, and take the opportunity to ask a classic question about the classification of quadrilaterals. The work is quick and without surprise for the teacher: some pupils say that it is false, because the diamond also has its four sides equal; others validate Socrates' answer because they only seek to distinguish the square from the rectangle, as in the worksheet below (Fig. 5.4).

Second Session

The setup here is different. As we want the pupils to work on mathematical aspects without having access to the text, this latter is not distributed to them. It is gradually projected by the teacher, who relies on it to lead the questioning. The session begins with an exchange in the form of questions and answers aimed at remembering the important points of the first session: we are working on a text from ancient Greece, written by Plato, in which three characters are dialoguing. Socrates is trying to convince Meno of something, but we don't know what yet. To do this, he questions a slave, starting by presenting him with a square whose side is two feet (in length) and the area four feet (of area). We draw Fig. 5.2b on the board.

Moment 1. Discovery of the Main Problem. Comparison of Pupils' Responses to That of the Slave

We project the last sentence of the first excerpt; we ask the pupils to propose to reformulate it, to say it with their own words.

> SOCRATES. And might there not be another figure twice the size of this, but of the same sort, with all its sides equal like this one ?

We quickly agree that the requested figure is a square. The formulation "twice" is however a bit ambiguous: the new square could be double in length or double in area. The term "twice the size" could suggest that it is rather "double in area", but this is not certain. Moreover, perhaps it is the same square that doubles both the area and the perimeter, in which case there is no need to distinguish.

> SOCRATES. And might there not be another figure twice the size of this, but of the same sort, with all its sides equal like this one ?
> BOY. Yes.
> SOCRATES. Then how many feet will it be?
> BOY. Eight.

Question: do these few lines help to lift the ambiguity of Socrates' question?

Yes: a square doubled in length would have a perimeter of 2×8 feet long, while a square doubled in area would be 2×4 feet in area. It is indeed a square doubled in area that we are talking about here.

Question: what would be your answer to Socrates' question? I will give you a minute, I just want to see what your ideas are, your proposals. Don't hesitate to draw figures.

We see that our goal is not to establish a research phase but rather to allow pupils to appropriate the question, even if—as we anticipate—they make the expected mistake and propose to double the side.

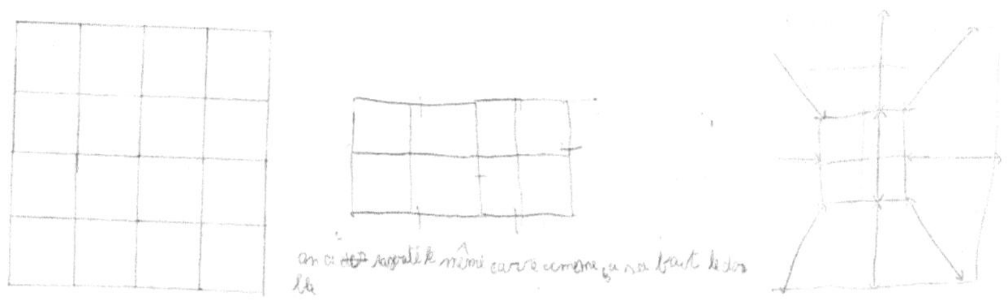

Fig. 5.5 How to double the area of the initial square? Three worksheets (English translation) "We added the same square so it makes the double." (original spelling)

In the classes, this phase was richer than we had anticipated. Of course, no pupil finds a correct solution, nothing can lead to it in this quick brainstorming phase. Here are the written traces of the research of three pairs of pupils (Fig. 5.5):

We see at the left the expected mistake; the figure being already gridded in unit areas, we could rely on it to quickly invalidate this proposal. Under the figure on the middle the pupils wrote "We added the same square so it makes the double"; we can refer the proposal back to the class for invalidation: indeed, we have produced a figure of double area, but it is not a square. The proposal on the right is probably to be read dynamically: the pupils indicate by double arrows a stretching of the initial square (in the centre), as if they were working on a touch screen. We can agree with the pupils that this proposal is not precise enough, for lack of knowing how to stretch to obtain exactly double the area.

The rest of the text allows us to compare the pupils' answers to that of the slave:

> SOCRATES. Come now, try and tell me how long will each side of that figure be. This one is two feet long: what will be the side of the other, which is double in size?
>
> BOY. Clearly, Socrates, double.

Moment 2. A Difficult Passage to Understand

We invite the pupils to read silently the rest of the text, which we project:

> SOCRATES. Do you observe, Meno, that I am not teaching the boy anything, but merely asking him each time? And now he supposes that he knows about the line required to make a figure of eight square feet; or do you not think he does?
>
> MENO. I do.
>
> SOCRATES. Well, does he know?
>
> MENO. Certainly not.
>
> SOCRATES. He just supposes it, from the double size required?
>
> MENO. Yes

We lead the pupils to clarify the following points: here Socrates stops talking to the slave and resumes the conversation with Meno; Socrates reformulates the slave's answer: the square of eight feet in area is formed, according to the slave, by doubling the side of the square of four square feet. Socrates and Meno seem to think that the slave is mistaken, or at least that there is a problem with his answer: there is something that the slave does not "know".

Pupils are told that the meaning of this aside between Socrates and Meno will be better understood after reading the rest of the text.

Moment 3. Socrates Invalidates the Slave's Answer

The rest of the text is distributed, reproduced here in Appendix 2. The sign [. . .] indicates a cut that does not need to be justified to the pupils. We have removed two pages where the slave, after proposing a 4-foot square, tries the one of 3 feet. The text is distributed on a sheet on which one or two blank spaces have been left.

> Instruction: I am giving you the rest of the text. As in the first session, the text is a bit difficult to read because the characters are talking about figures that are not drawn. Take 5 minutes to read this text in silence. I ask you to use the space left on the sheet to draw the figures that Socrates talks about. When you have finished:
>
> - We will look at the figures you have proposed.
> - I will ask you—without rereading the text—to explain to me what is happening, in your own words.

The worksheets distributed to the pupils were slightly different depending on the classes.

This transition from dimension 1 to dimension 2 is perfectly captured in the top worksheet (Fig. 5.6), although the numerical values suggest a confusion between lengths and areas (the side that is doubled is 2 feet long, it is the initial square that is 4 square feet of area). Even if the bottom worksheet (Fig. 5.6) shows surfaces from the start, it is the lengths that are indicated in the top part, the areas only appearing in the bottom part. Failing to have appropriated the coding "foot long"/"square foot", the pupil distinguishes the two by the placement of the coding on the figure, in a fairly clear way, moreover.

The work was more guided in level 6 class: in the only area to be completed by a diagram, the initial square was already given. Here are three answers:

Only the figure on the left corresponds to what the text describes. Overall, the productions of the 6-graders were more homogeneous and showed more correct figures than in grade 5, where the work was less guided. However, the process of going from the given text to producing figures led the pupils of class 5 to try to indicate the lengths and areas on their figures, whereas, as the answers of Fig. 5.7 illustrate, the provision of the initial square without coding (other than the grid in unit square) seems to inhibit work on numerical information—almost absent in the worksheets. We recommend combining the format used in grade 6—guided by the provision of an initial square to be completed—and the invitation to indicate the measurements on the figure by indicating at least the lengths on the given square.

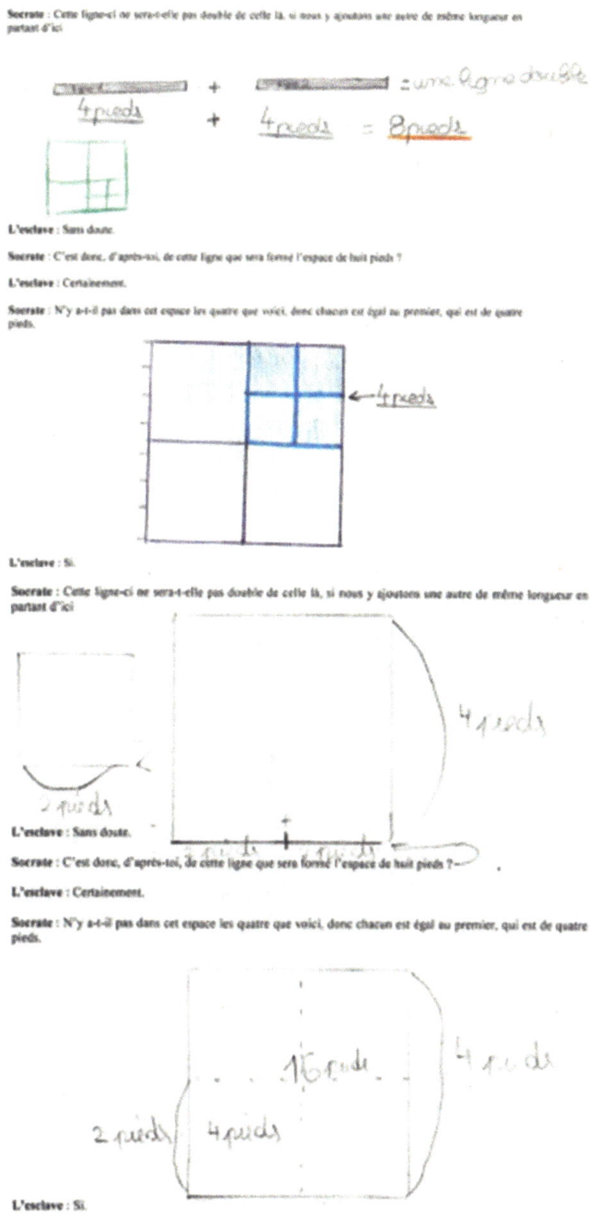

Fig. 5.6 Text to be completed with figures. Two worksheets

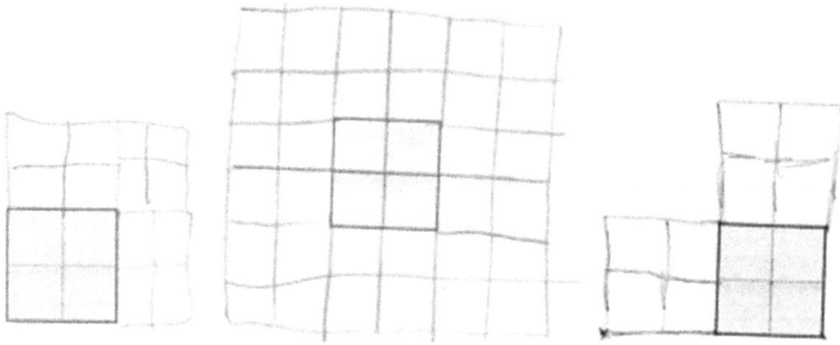

Fig. 5.7 Three figures attempting to illustrate the doubling of the side

That being said, whatever the diversity of the pupils' proposals, consensus is easily reached during the correction: the figure mentioned in the text is the one on the left (Fig. 5.7). Once the correct figure is in front of them, the pupils have no difficulty in reformulating Socrates' argument—even if it means going to the board to rely on the figure. It is agreed that the slave's proposal does not answer the question: by doubling the side of the square, we went from 4 square feet to 16 square feet; we quadrupled, not doubled.

Moment 4. Return to the Aside Between Socrates and Meno
We return to the aside between Socrates and Meno.

Questions:

- Did Socrates and Meno know that the slave's answer was false?
- Can you explain "certainly not"?

The first question poses no difficulty.

The second is more arduous. The identification of verbs invites us to distinguish two lexical fields: on one side what one "knows"; on the other what one imagines ("he supposes he knows"). Socrates and Meno observe that the slave is perfectly convinced of the correctness of his answer, even though it is false. They reserve the verb "to know" for justified assent to true propositions; in this sense, the slave "does not know".

These more philosophical aspects do not seem totally inaccessible to us as long as we illustrate them. We can take an example: if you ask a child "what do 2 times 3 make" and he answers "5", will you say that he "knows" that $2 \times 3 = 5$ or that he "believes" or "thinks he knows" or "imagines" that $2 \times 3 = 5$?

Third Session

The end of the text consists of Socrates' masterful exposition, which describes a correct construction, and justifies it using the initial half-square (an isosceles right triangle, therefore) as a unit of area allowing to measure the initial square (of measure 2) and the new square (of measure 4). We made two choices. Firstly, the work of constructing the figure from the indications of the text did not seem to bring new elements; the text is very ambiguous when it mentions a "line drawn from corner to corner". The excerpt reproduced in Appendix 3 is distributed to the pupils; the teacher reads it aloud while drawing the elements of the figure on the board as he goes along. Figure 5.1d (see the introduction of this chapter) remains on the board throughout the session.

Secondly, we chose not to give the pupils Socrates' justification, but to let them search for their answers to the question on which we stop reading: "how large this space?" (we will invite to reformulate: "what is the area of this surface?").

Moment 1. Validation, by the Pupils, of Socrates' Proposition
By asking the pupils to recall what the problem posed to the slave is, we agree that there are two points to check:

- Is the "tilted" quadrilateral that appears in the centre of Fig. 5.1d really a square?
- If yes, is this square double the area of the initial square?

For the first question, we ask the pupils to write on their notebook the instruments they would like to use to answer it. The objective is to revisit the instrumented characterisation of the square; in particular, the fact that checking the equality of lengths is not enough to characterise the square among quadrilaterals. This is also an opportunity to remind that we can have "tilted" right angles, that is to say, elsewhere than between a horizontal and a vertical. It goes without saying that, in the course of schooling, we will be able to demonstrate that it is indeed a square, using for example the case of equality of Side-Angle-Side triangles (Euclid's *Elements*, Book I prop. 4 (Euclid, 1956, pp. 247–250)).

For the second question, we wanted to see a wide variety of procedures emerge, and get pupils to think *before* doing, hence the reintroduction of the "order form" device. For this, we had planned three "toolboxes" (exercise sheets with figures and a list of authorised instruments) to answer the question. The content of each of the three is presented to the pupils, and a few minutes are given to the pairs to decide which of the three they want to order. After experimentation, it seems that this device is not necessary to bring out many procedures, all correct.

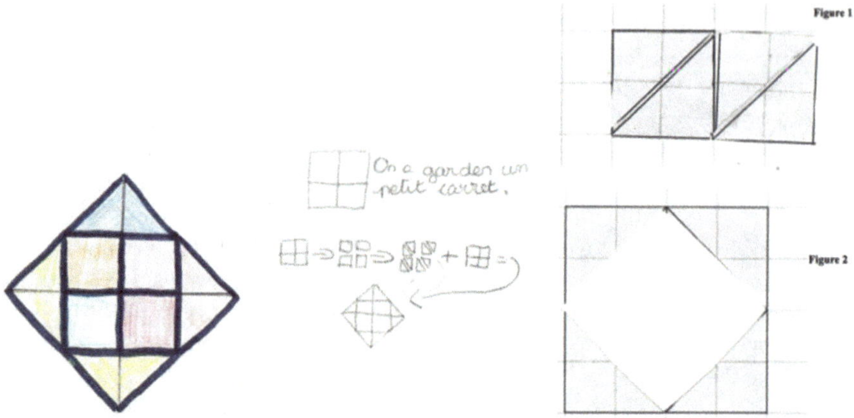

Fig. 5.8 Two solutions by cutting and reassembling (English translation) "Does the square have an area of 8 feet? Is it double the initial square?" "We can keep a small square"

We offered the pupils a first toolbox: a sheet with Fig. 5.1d, the graduated ruler and the calculator; we took the *cm* as a unit of length rather than the foot. This framework invited to measure the side of the tilted square with the ruler, and to use the formula of the area of the square. The result obtained is probably not equal to 8, due to measurement uncertainties. However, it is plausible that the tilted square has an area of 8 cm^2. In the classes, this solution attracted the attention of very few pupils. Most of the pairs who ordered this toolbox did not know how to use it, which means they had not identified the suggested numerical procedure. All the more reason for the teacher to mention it during the collective review phase.

The second toolbox included Fig. 5.1a, d, scissors and glue. Two families of correct procedures are observed:

In class 6 (on the right in Fig. 5.8), Fig. 5.1a was given at the top (on a grid background) and Fig. 5.1d at the bottom. Here, the pupil cut out the tilted square from the bottom figure, re-cut it into four right triangles with which he reconstituted two worksheets of the initial square, one glued back onto the square given at the top. It is noticeable that he did not choose to join the squares to make a rectangle, perhaps because in session 2 this rectangle had been declared incorrect. This difficulty in choosing the "shape" to appear at the top also probably explains the few incorrect procedures observed. Indeed, some pupils tried to reconstitute at the top a figure that had been encountered during the three sessions, even if it did not double the initial area. We find the well-known difficulty in abandoning—when relevant—the "shape" aspect to retain only the "size" aspect.

In class 5 (on the left in Fig. 5.8), two worksheets of the initial square were given. The pupil made the tilted square from the two given squares, and indicated his "construction" programme. This beautiful solution was proposed by several pairs.

The third toolbox contained Fig. 5.1d and nothing else. The pupils justified the fact that the tilted square is indeed double—in area—of the initial square by

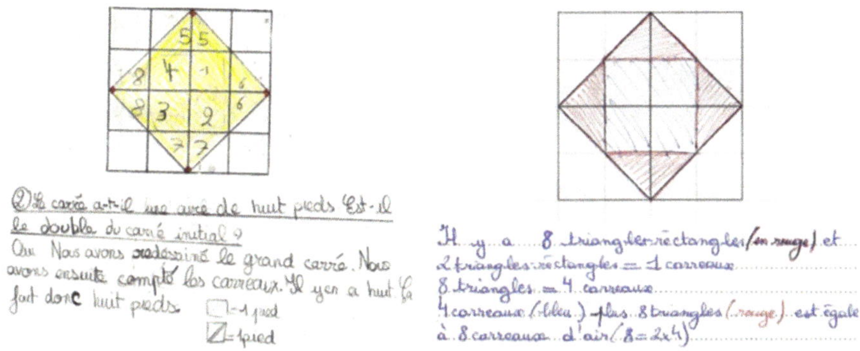

Fig. 5.9 Two solutions by coding on the figure (English translation): Left worksheet: “Does the square have an area of eight feet. Is it double the initial square? Yes. We have redrawn the large square. We then counted the squares. There are eight. So that makes eight feet.” Right worksheet: “There are 8 right triangles (in red) and 2 right triangles = 1 square. 8 triangles = 4 squares. 4 squares (blue) plus 8 triangles (red) equals 8 squares of area (8 = 2 × 4).” (Original spelling)

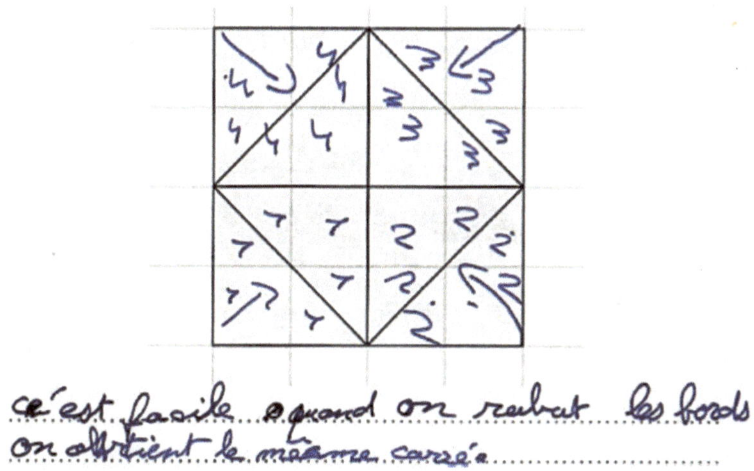

Fig. 5.10 A solution by folding (English translation) “It’s easy, when you fold the edges you get the same square”

comparisons of areas based on colour designations of parts of the figure. Here (Fig. 5.9) are two correct productions, on the left in class 5, on the right in class 6.

In class 6, a pair proposed a correct solution that we had not anticipated (Fig. 5.10).

The folds allow us to describe the tilted square as half—in area—of the large square. The large one being 16 square feet, the tilted square is indeed 8 square feet. We also approach Socrates’ solution, which consists of taking the right-angled triangle as a standard of area.

Moment 2. Quick Review of the General Meaning of the Text

We thought it was a pity not to revisit the movement and the general meaning of the text, especially since we had told the pupils that passages that were not understandable at first reading would be clarified by elements discovered later. To enter this final phase, we project again the first lines of the dialogue (Appendices 2, 3). The pupils recognise the starting point of the work of the three sessions. We then ask them to tell, in their own words, what the dialogue is about. To invite narration, we can suggest the use of the words "first", "then", "next", even if the pupils prefer "after". This work of remembrance and reordering of episodes is quite difficult, but the mathematical thread is sufficiently present in the pupils' minds for an exchange with the teacher to stabilise the following elements: Socrates seeks to convince Meno of something; to do this, he asks Meno if he can "use" one of his slaves. Socrates checks that the slave knows what a square is and knows the area of a square with a side 2 feet long: it's 4 square feet. Socrates asks the slave to find a square double in area, therefore 8 square feet. The slave finds the question easy, and proposes to double the side of the given square. Socrates shows him that he is wrong—Socrates and Meno knew this from the start—the square with double side is quadruple in area, not double. Socrates then draws a new figure, which is indeed the square sought.

It remains to imagine what Socrates seeks to prove to Meno through this mathematical digression. We ask two questions to the pupils; they are written on the board; we announce that we leave a few minutes for reflection:

- Is the goal of Socrates and Meno to mock the slave?
- Does the dialogue between Socrates and Meno focus on squares and areas? If not, what on?

The pupils understand well that the goal is neither to mock nor to humiliate the slave. Socrates' attitude towards him is professorial, and the pupils see in the time he spends dialoguing with him the sign of a benevolent attention. As for the second question, let's say right away that the pupils do not have enough information to go beyond the stage of conjecture. Remember that we are not aiming here for an initiation into philosophy but a work on the formulation of reading hypotheses supported by identifiable elements in the text. The teacher has some leeway here, for example by playing on cuts in the text. Thus, the fundamental philosophical passage given in Appendix 5 was read in grade 5 during session 2, while it had been cut from the text read in class level 6 (Appendix 3).

The second question invites us to distinguish, on the one hand, the dialogue between Socrates and the slave—which indeed focuses on squares and areas—, on the other hand, the dialogue between Socrates and Meno. At the very least—that is to say without the passage of Appendix 5—we can lead the pupils to propose the following points: Socrates and Meno are interested in the way we learn, how we solve problems. They lead the slave to propose a false answer, but it is not to mock him; when the problem is difficult, one can make a mistake first but find or understand afterwards. The text in Appendix 3 allows a more detailed reading: often we imagine knowing things that are yet false, and we must realise this in order to then learn. Without instruction one can find oneself either in ignorance (we do not know) or in illusion (we believe we know, but we hold as true what is not). In the minimal version, the pupils retain from the end of the text that even a slave, someone who has received no mathematical education, can understand if it is well explained to him. Of course, we are not saying that this passage from *Meno* is *for Plato* a plea for a democratic education. This reading hypothesis is however reasonable given the passages read, and, in both classes, it is the one that was proposed by the pupils.

Assessment and Suggestions

From the point of view of the teachers who designed and implemented the sequence, the set of three sessions leads to rich work both in mathematics and in text comprehension. The enrolment of pupils in the work is however more difficult in the first session, some pupils being quickly discouraged by the initial reading of a text so obscure to them.

Session 1 also presents a second difficulty, that of the indiscriminate use by Socrates of the term "foot" to designate the unit of length and the associated area unit. It is important that at the time of the assessment on this point the teacher is very explicit about the importance of distinguishing the two. In both classes, enough pupils had grasped this point when they had to illustrate and code the figures so that the teacher could rely on them in the collective phase. To allow each pupil to appropriate the distinction between "feet of length" and "feet of area" (or "square feet") after it has been validated by the teacher, we propose to add a small activity to the course of session 2: ask pupils to code with different colours in the text the places where we should have "feet of length" and those where we should have "feet of area".

The two following sessions are more guided and more focused on the mathematical problem. They proceed as anticipated: pupils' ability to engage with the problems, production of expected correct and incorrect answers. The pupils' productions in session 3 showed a greater variety of correct procedures than we had anticipated. On the other hand, we were surprised by the rarity of the use of the numerical framework, even in grade 6, where the work by cutting, pasting, and tiling tends to fade.

In grade 5, a small questionnaire was given to the pupils one week after the implementation of the sequence:

> Do you agree with the following statements?
>
> - "To double the area of a figure, it is enough to double all its lengths."
> - "When a quadrilateral has its four sides of equal lengths, it is necessarily a square."

As expected, the vast majority of pupils invalidate the second statement and cite the case of the diamond.

For the first question, a few pupils have retained that it is false and can cite the example studied in the *Meno*. However, the vast majority of pupils respond that they agree with the first sentence, with varied arguments, either showing a difficulty in articulating reasoning about lengths and reasoning about areas, or showing uncertainty about the meaning of "doubling" (or the means of controlling the doubling). This can be seen in the following three answers (Fig. 5.11).

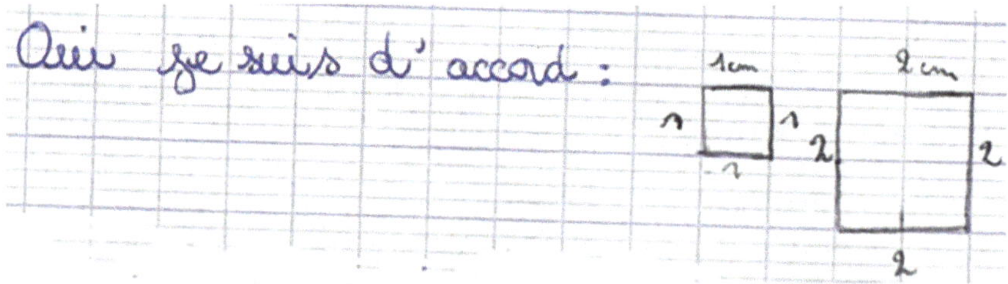

Worksheet #1

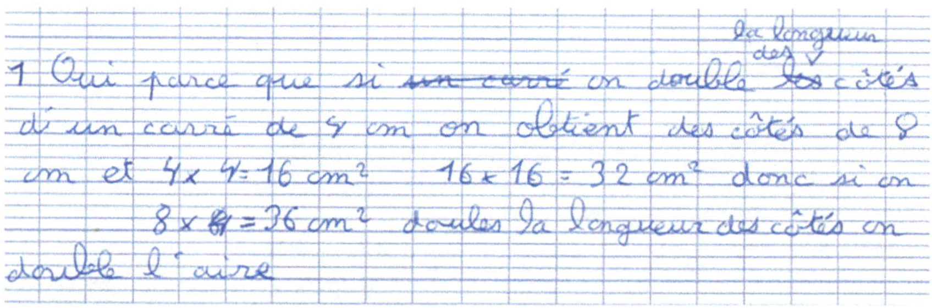

Worksheet #2

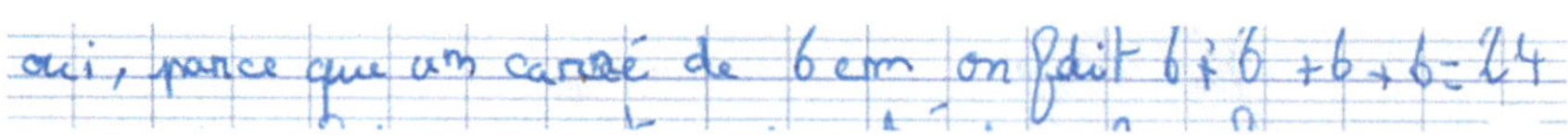

Worksheet #3

Fig. 5.11 Three incorrect answers. (English translation) Worksheet 1: "Yes, I agree."; Worksheet 2: "Yes because if we double the length of the sides of a square of 4 cm we get sides of 8 cm and $4 \times 4 = 16$ cm^2, $16 + 16 = 32$ cm^2, $8 \times 4 = 36$ cm^2 so if we double the length of the sides we double the area."; Worksheet 3 "Yes, because a square of 6 cm we do $6+6+6+6 = 24$"

Here again, the presence of correct and well-justified answers in some pupil worksheets allows for a pooling based on exchanges of arguments among pupils. The summary should explicitly highlight, as something to remember, the fact that doubling the lengths does not double the area; at least not always. This can be linked to work on standard units and conversions. Thus, while there are indeed 10 decimetres in 1 metre, there are 100 dm^2 (and not 10) in 1 m^2: tenfold the side does not tenfold the area. This explains the splitting of the columns in the conversion table of areas measured in the metric system.

Acknowledgments This chapter is the result of the work carried out within the M.:A.T.H. group of the IREMS of Paris (France) by Renaud Chorlay (INSPE of Paris and LDAR), Alexis Gautreau (Rodin high school, Paris 13th) and Dominique Heguiaphal (Arago elementary school, Paris 13th).

Appendix 1: Concepts and Skills Addressed in This Chapter

Concepts Covered

- Comparing, classifying and ordering surfaces according to their areas without resorting to measurement.
- Differentiating between the area and perimeter of a surface.
- Determining the measurement of a surface area from a simple tiling or using a formula.
- Recognising, naming, comparing, verifying, describing simple or complex figures.
- Initial characterisations of specific quadrilaterals.
- Enlargement or reduction of a figure.

Skills Developed

- **Search and research skills**: Extract and organise the information necessary for problem-solving from various sources: texts, drawings, diagrams. Engage in a process, observe, question, manipulate, experiment, hypothesise, by developing reasoning adapted to a new situation. Attempt, try several problem-solving paths.
- **Reasoning**: Progress collectively in an investigation by taking into account the viewpoint of others. Justify assertions and seek the validity of the information at hand.

(continued)

- **Mathematical communication skills**: Gradually use adequate vocabulary and/or adapted notations to describe a situation, present an argument. Explain one's approach or reasoning, understand the explanations of another and argue in the exchange.
- **Mathematical representation ability**: Use tools to represent a problem: drawings, diagrams. Analyse a plane figure from different aspects.
- **Modeling**: Use geometric properties to recognise objects.

Appendix 2: Plato (1924, pp. 303–307)

[...]

MENO. I assure you, Socrates, that was not my intention; I only spoke from habit. But if you can somehow prove to me that it is as you say, pray do so.

SOCRATES. It is no easy matter, but still I am willing to try my best for your sake. Just call one of your own troop of attendants there, whichever one you please, that he may serve for my demonstration.

MEN. Certainly. You, I say, come here.

SOC. He is a Greek, I suppose, and speaks Greek?

MEN. Oh yes, to be sure – born in the house.

SOC. Now observe closely whether he strikes you as recollecting or as learning from me.

MEN. I will.

SOC. Tell me, boy, do you know that a square figure is like this?

BOY. I do.

BOY. Now, a square figure has these lines, four in number, all equal?

BOY. Certainly.

SOC. And these, drawn through the middle, are equal too, are they not?

BOY. Yes.

SOC. And a figure of this sort may be larger or smaller?

BOY. To be sure.

SOC. Now if this side were two feet and that also two, how many feet would the whole be? Or let me put it thus: if one way it were two feet, and only one foot the other, of course the space would be two feet taken once?

BOY. Yes.

SOC. But as it is two feet also on that side, it must be twice two feet?

BOY. It is.

SOC. Then the space is twice two feet?

BOY. Yes.

SOC. Well, how many are twice two feet? Count and tell me.

Boy. Four, Socrates.

Soc. And might there not be another figure twice the size of this, but of the same sort, with all its sides equal like this one?

Boy. Yes.

Soc. Then how many feet will it be?

Boy. Eight.

Appendix 3: Plato (1924, pp. 309–313)

[...]

Soc. Well, this line is doubled, if we add here another of the same length?

Boy. Certainly.

Soc. And you say we shall get our eight-foot space from four lines of this length?

Boy. Yes.

Soc. Then let us describe the square, drawing four equal lines of that length. This will be what you say is the eight-foot figure, will it not?

Boy. Certainly.

Soc. And here, contained in it, have we not four squares, each of which is equal to this space of four feet?

Boy. Yes.

Soc. Then how large is the whole? Four times that space, is it not?

Boy. It must be.

Soc. And is four times equal to double?

Boy. No, to be sure.

Soc. But how much is it?

Boy. Fourfold.

Soc. Thus, from the double-sized line, boy, we get a space, not of double, but of fourfold size.

Boy. That is true.

Soc. And if it is four times four it is sixteen, is it not?

Boy. Yes.

Soc. What line will give us a space of eight feet? This one gives us a fourfold space, does it not?

Boy. It does.

Soc. And a space of four feet is made from this line of half the length?

Boy. Yes.

Soc. Very well; and is not a space of eight feet double the size of this one, and half the size of this other?

Boy. Yes.

Soc. Will it not be made from a line longer than the one of these, and shorter than the other?

[...]

Soc. But from what line shall we get it? Try and tell us exactly; and if you would rather not reckon it out, just show what line it is.

Boy. Well, on my word, Socrates, I for one do not know.

Appendix 4: Plato (1924, pp. 315–317)

[. . .]

Soc. [. . .] Tell me, boy: here we have a square of four feet, have we not? You understand?

Boy. Yes.

Soc. And here we add another square equal to it?

Boy. Yes.

Soc. And here a third, equal to either of them?

Boy. Yes.

Soc. Now shall we fill up this vacant space in the corner?

Boy. By all means.

Soc. So here we must have four equal spaces?

Boy. Yes.

Soc. Well now, how many times larger is this whole space than this other?

Boy. Four times.

Soc. But it was to have been only twice, you remember?

Boy. To be sure.

Soc. And does this line, drawn from corner to corner, cut in two each of these spaces?

Boy. Yes.

Soc. And have we here four equal lines containing this space?

Boy. We have.

Soc. Now consider how large this space is.

Appendix 5: Plato (1924, pp. 313–315)

[. . .]

Boy. Well, upon my word, Socrates, I for one do not know.

Soc. There now, Meno, do you observe what progress he has already made in his recollection? At first he did not know what is the line that forms the figure of eight feet, and he does not know even now: but at any rate he thought he knew then, and confidently answered as though he knew, and was aware of no difficulty; whereas now he feels the difficulty he is in, and besides not knowing does not think he knows.

Men. That is true.

Soc. And is he not better off in respect of the matter which he did not know?

MEN. I think that too is so.

SOC. Now, by causing him to doubt and giving him the torpedo's shock, have we done him any harm?

MEN. I think not.

SOC. And we have certainly given him some assistance, it would seem, towards finding out the truth of the matter: for now he will push on in the search gladly, as lacking knowledge; whereas then he would have been only too ready to suppose he was right in saying, before any number of people any number of times, that the double space must have a line of double the length for its side.

MEN. It seems so.

SOC. Now do you imagine he would have attempted to inquire or learn what he thought he knew, when he did not know it, until he had been reduced to the perplexity of realising that he did not know, and had felt a craving to know?

MEN. I think not, Socrates.

SOC. Then the torpedo's shock was of advantage to him?

MEN. I think so.

SOC. Now you should note how, as a result of this perplexity, he will go on and discover something by joint inquiry with me, while I merely ask questions and do not teach him; and be on the watch to see if at any point you find me teaching him or expounding to him, instead of questioning him on his opinions. Tell me, boy: here we have a square of four feet, have we not? You understand?

References

Burton, D. (2011). *The history of mathematics: An introduction*. Mc Graw Hill.

Cèbe, S., & Goigoux, R. (2009). *Lector & Lectrix. Apprendre à comprendre les textes. CM1-CM2-6e-Segpa*. Retz.

Euclid. (1956). *The thirteen books of Euclid's elements*. Dover Publications.

Netz, R. (2003). *The shaping of deduction in Greek mathematics: A study in cognitive history*. Cambridge University Press.

Plato. (1924). *Laches, Protagoras, Meno, Euthydemus*. William Heinemann & G.P. Putnam's son. https://archive.org/details/lachesprotagoras0000plat/page/303/mode/2up

Renaud Chorlay is a researcher in mathematics education at the LDAR research center in Paris, and a teacher educator in the Faculty of Education of Sorbonne University (INSPE de Paris). His main research interests are, on the one hand, mathematical argumentation, and, on the other hand, connections between history of mathematics and mathematics education.

Chapter 6
1793, The Revolution of Time

Frédéric de Ligt

> *It was at this moment when we did not fear doing violence to all received ideas, to all established habits, that the project to renew the system of weights and measures and to change the calendar was executed. The taste for regularity and the contempt for obstacles were to signal a revolution that was both philosophical and political.*
>
> Adolphe Thiers (1866, p. 696)

Abstract In this chapter, the IREM&S group from Poitiers is interested in the French Revolution during which it was a question of setting up a new calendar and a new division of the day. Goodbye to the 60 minute hour! The decimal system must be imposed at the same time as the desired unification of weights and measures with the metric system. This is not without posing conversion problems, the same ones which inform the work proposed here to ten and eleven year old pupils based on legislative texts of the time or the Republican calendar put in place at the time.

Keywords Decimal system · French Revolution · Decimal time · Proportionality · Republican calendar

The decimal system, fractions, conversions, information gathering, introduce many difficulties encountered by young pupils. Yet the stakes of succeeding in these learnings are important. A very short and little-known episode of the French Revolution saw the implementation of a new calendar and a new division of the day. We proposed to 10 and 11 year old pupils to work, from historical documents,

---

F. de Ligt (✉)
IREM&S de Poitiers, Poitiers, France
e-mail: frederic.de-ligt@ac-poitiers.fr

M. Moyon et al. (eds.), *Teaching Mathematics Through Historically-Based Activities*, History of Mathematics Education,
https://doi.org/10.1007/978-3-032-16990-7_6

on this other division of time. The numerous conversions necessary to make the correspondence with our usual fractioning of durations naturally mobilise this knowledge.

The magnitude studied, the motivations that led to this "revolution of time", the personality of the protagonists who participated in this scientific adventure during this burning period of the history of France, all this context should convince pupils that mathematics answers questions rooted in the lives of men, to make them discover the existence of the history of sciences and finally to make them want to better understand and master the mathematical tools useful for solving these questions.

Historical Context

The Unification of Weights and Measures Built According to the Decimal Division

King Louis XVI, in a difficult political and economic context, decided on 4 May 1789 to convene the *États Généraux*, it was an assembly made up of the three orders: the clergy, the nobility and the third estate, in order to decide on the reforms to be undertaken to remedy the situation.

The unification of weights and measures was widely demanded by the three orders in the grievance books of 1789, which prepared the *États Généraux*. The great fluctuation, over time and in different places of the territory, of the value of the same measure, the diversity of units and the inevitable frauds exceeded the population (Zupko, 1990). Moreover, this hindered the smooth running of trade and industry, but also the development of science (Dowson, 1891). For example, in the north of France, there were eighteen kinds of ells ranging between 62 and 84 centimetres!

"There is an infinity of different measures among the lords. It is requested that all measures be reduced to those of the king", *Grievance book*, Quimper (Bretagne). For unity to be achieved: "Each province would have to sacrifice part of its old customs and practices for the benefit of having a single law", Grievance book, Angers (Pays de Loire).

These grievances can be summarised by the formula "One king, one law; one weight and one measure". The night of 4 August 1789 abolished privileges and, in particular, that of calibration, which is no longer a prerogative of the King. A rational reform is then possible. It is about finding in nature, according to Talleyrand's ideas, a universal and invariable measure, reproducible and verifiable everywhere and always.

On 8 May 1790, the National Assembly, on the proposal of Talleyrand, entrusted a commission of the Academy of Sciences composed of Condorcet, Lagrange, Laplace, Monge, Borda, Tillet and Lavoisier with the task of producing a report on the unification of the measurement system. On 26 March 1791, on the recommendation of Condorcet, the quarter of the terrestrial meridian was chosen as the universal and natural base of the unit of measure. On 26 March 1791, the following decree was adopted:

> The national assembly, considering that to achieve the uniformity of weights and measures, in accordance with its decree of 8 May 1790, it is necessary to fix a natural and invariable unit of measure, and that the only way to extend this uniformity to foreign nations, and to encourage them to agree on a same system of measure, is to choose a unit, which, in its determination, contains nothing arbitrary, nor particular to the situation of any people, on the globe; considering that more than the unit proposed in the notice of the Academy of Sciences of 19 March of this year, meets all the conditions, has decreed and decrees that it adopts the magnitude of a quarter of the terrestrial meridian as the basis for the new system of measurements; that, as a result, the necessary operations to determine this base, as they are indicated in the notice of the Academy, and in particular the measurement of an arc of the meridian from Dunkirk to Barcelona, will be promptly executed; that, as a result, the King will charge the Academy of Sciences to appoint commissioners who will promptly deal with these operations and will consult with Spain for those that must be done on its territory. (Assemblée Nationale Française, 1791)

The National Assembly demanded in December 1792 that the Academy set provisional standards. The length of the quarter meridian that had been measured by Cassini and Lacaille is therefore taken as the basis of the unit, while a team of scientists measures it with more precision. This will be the adventure of measuring the meridian between Dunkirk and Barcelona by Delambre and Méchain, who will be in charge of the triangulation. In May 1793, the mission delivers its report. Borda in particular proposes to call "metre" from the Greek *metron* (measure) the ten-millionth part of the quarter of the terrestrial meridian. This new unit will follow the decimal division as recommended from 1790. However this idea of decimal division in measurements is not new, it dates back to antiquity. Simon Stevin, as early as the sixteenth century, in his work *La Disme*, had perfected this system to make it operational with the use of decimal fractions and the comma. Thus, shortly before the Revolution began, in the *Encyclopaedia*, in the article "Decimal" written by D'Alembert, one can read:

> It would be very desirable that all divisions, for example of the pound, the penny, of the yard, the day, the hour, etc., were from 10 to 10; this division would make the calculation much easier and more convenient, and would be much preferable to the arbitrary division of the pound into 20 pennies, the penny into 12 deniers, the day into 24 hours, the hour into 60 minutes, etc. (D'Alembert, 1754, pp. 669b–670a)

The law was adopted on 1st August 1793 (Assemblée Nationale Française, 1793).

This unification also has a symbolic scope, as it contributes to the construction of national cohesion. Thus the first article of the law of 18 Germinal of the year III (7 April 1795) calls on citizens to demonstrate

> a proof of attachment to the unity and indivisibility of the Republic by using the new measures in calculations and commercial transactions (Assemblée Nationale Française, 1795).

The Republican Calendar and the Decimal Division of the Day

How does the reform of the calendar and the division of the day fit into this great movement of questioning the measurements hitherto in force?

On 20 September 1792, the army of the Revolution defeated the Prussian, Austrian and French royalist emigrants at Valmy, which saved the French Revolution; on 21 September, meeting for the first time, the deputies of the National Assembly abolished the monarchy; on 22 September 1792, they founded the Republic and decided that all public acts would henceforth be dated from year I of the Republic. The Assembly asked its Public Instruction Committee, chaired by Charles-Gilbert Romme, to present "as soon as possible a project on the benefits that the agreement of its republican era with the vulgar era should bring to France" (Convention Nationale Française, 1792, p. 299).

A commission was then formed around Romme (deputy of Riom in Auvergne), Claude-Joseph Ferry (deputy of the Ardennes and professor of engineering at the military school of Mézières) and Charles-François Dupuis (deputy of Seine-et-Oise, astronomer, mathematician, jurist and specialist in ancient calendars) with the mission of proposing a new calendar that would no longer refer to royalty or Christianity and whose division of the year would use the decimal system. The astronomers and mathematicians Joseph-Louis Lagrange and Joseph-Jérôme Lefrançois de Lalande, the mathematician Gaspard Monge, Louis-Bernard Guyton-Morveau (deputy of Côte-d'Or and chemist) and Alexandre-Guy Pingré, a scholar, joined this commission.

After nine months of work, Romme, appointed rapporteur of the commission by the Public Instruction Committee, presented a calendar project to the National Assembly on 20 September 1793. After a few minor modifications, a decree was adopted on 5 October 1793 (Convention Nationale Française, 1793a), and came into force the next day, with a retroactive application date of 1st Vendémiaire year I (22 September 1792), the day of the proclamation of the Republic and the first day of the true autumn equinox in Paris. On 24 October, the nomenclature of the calendar proposed by Fabre d'Églantine was voted and a month later, on 24 November 1793, a revision of the decree brought clarifications, notably on the names given to the months and days, but also on the division of time. The historian Jules Michelet was able to say on this subject: "For the first time in this world, man had the true measure of time" (Michelet, 1853, p. 182).

In this decree of 5 October 1793, with the Republican calendar, a new division of the day based on the decimal system was instituted.

> XI. The day, from midnight to midnight, is divided into ten parts or hours, each part into ten others, and so on until the smallest commensurable portion of duration. The hundredth part of the hour is called *decimal minute*; the hundredth part of the minute is called *decimal second*. This article will only be mandatory for public acts, from 1st Vendémiaire, year three of the Republic. (Convention Nationale Française, 1793b)

This decimal division of the day then encountered many obstacles in its application and a great reluctance in the population (Vera, 2008, pp. 35–36). The decimal hour was suspended by article 22 of the decree on weights and measures of 18 Germinal year III (7 April 1795):

> The provision of the law of 4 Frimaire year II, which makes mandatory the use of the decimal division of the day and its parts, is indefinitely suspended (Assemblée Nationale Française, 1795).

As for the Republican calendar, the fact that the French have mostly continued to think of time according to the old calendar led to the decision of 22 Fructidor year XIII (9 September 1805) taken by Napoleon, who then signed the *senatus consultum* establishing the return to the Gregorian calendar on 1st January 1806.

The Organisation of the Republican Calendar

As with the Julian and Gregorian calendars, Romme calls on astronomers to undertake a reform of the calendar. A common point can be established between the Commission of Weights and Measures and the Calendar Commission. Politicians supervise the scientists, as the objectives of each are often different. The Commission begins by agreeing on the most precise possible duration of the year: 365 days, 5 hours, 48 minutes and 49 seconds. The year of the Republican calendar is still divided into twelve months, twelve as the number of times the Moon interposes between our planet and the Sun. Each month now has a fixed duration of thirty days. Each month is itself divided into 3 decades of 10 days to be based on the decimal system. Five to six complementary days (the *sans-culottides*) depending on the years, are added at the end of the year to keep it aligned with the tropical year. The four-year period that then opens is called a *franciade*. This sixth added day is called the day of the Revolution and is considered a national holiday.

Differences of views between the famous astronomer Lalande and the members of the Commission opposed them on a particular point of the decree of 5 October 1793 in its articles II (the vulgar era is abolished for civil uses) and III (the beginning of each year is fixed at midnight, starting the day when the true autumn equinox falls for the Paris observatory). Indeed, Lalande points out that starting the year on the day of the true autumn equinox prevents having fixed intercalation rules, rules that determine common years of 365 days and years of 366 days. The duration of the year would then be fixed by observation and would no longer be predictable. Romme writes in his report:

> It is only by intercalating sometimes days, sometimes months, at regular intervals, that the coincidence of the civil year with the celestial movements and the seasons was brought about. All these intercalations, made without fixed rules, temporarily repaired the effects of a faulty computation and left the original cause to persist (Romme, 1793).

There is a certain bad faith or a form of blindness in speaking thus of the Gregorian calendar, which can undoubtedly be explained by the political context of these years 1792 and 1793. Lalande objected to him: "The form of the Gregorian calendar is of sufficient accuracy; [...] 28 days would have to be removed in 36 centuries, instead of the 27 that are actually removed" (Froeschlé, 1996, p. 309). To which Lagrange replied, "that they did not want this rule, because it was Roman, and that they wanted a new form attached only to the day of the equinox" (Froeschlé, 1996, p. 312). As for Laplace, he concedes that a rule would be practical, but that the decree having passed, he prefers to say nothing more. However, through the force of argument, Romme finally lets himself be convinced that a limited reform is needed, which would allow astronomers to easily establish almanacs. On 18 April 1795, an assembly of scholars meets to find an amendment to article III of the decree of 5 October 1793; Delambre is tasked with drafting a report on the question. Delambre takes as the new duration of the year the value found by Lalande, 365 days 5 h 48 min 48 seconds, which differs by one second from the previous value. He makes various proposals for intercalation and finally Romme chooses the classic Gregorian intercalation, corrected by the removal of the sextile 4000 and its multiples, as calculated by Lagrange. He can thus continue to say that it is a different calendar! The Committee of Public Instruction approves this report, but the events of Prairial and the tragic end of Romme leave the calendar unfinished. Delambre writes on this subject:

> We could have found another reporter in the Committee of Public Instruction, but the one we approached did not dare propose any reform, for fear that they would completely abolish this calendar, instead of correcting it. (Froeschlé, 1996, p. 316)

The contribution of the poet Fabre d'Églantine and the gardener of the Jardin des Plantes of the National Museum of Natural History, André Thouin, can be cited for the development of the names that appear in this new calendar (Perovic, 2012, pp. 117–124).

The autumn months end in *-aire*:

- Vendémiaire (22 September to 21 October), grape harvest period.
- Brumaire (22 October to 20 November), period of mists and fogs.
- Frimaire (21 November to 20 December), period of cold weather (frosts).

The winter months end in *-ôse*:

- Nivôse (21 December to 19 January), period of snow.
- Pluviôse (20 January to 18 February), period of rain.
- Ventôse (19 February to 20 March), period of winds.

The spring months end in *-al*:

- Germinal (21 March to 19 April), period of germination.
- Floréal (20 April to 19 May), period of the blossoming of flowers.
- Prairial (20 May to 18 June), period of meadow harvests.

The summer months end in *-idor*:

- Messidor (19 June to 18 July), period of harvests.
- Thermidor (19 July to 17 August), period of heat.
- Fructidor (18 August to 16 September), period of fruits.

Each of the days was chronologically identified by an order: Primidi, Duodi, Tridi, Quartidi, Quintidi, Sextidi, Septidi, Octidi, Nonidi, Décadi.

To replace the names of saints, the days were assigned names of animals, plants, tools or agricultural products. Universal in its division, this calendar was, however, geographically marked by the choice of introduced names.

Here is the comment of Fabre d'Églantine in his Report made to the National Convention, in the session of the 3rd of the second month of the second year of the French Republic, on behalf of the Commission in charge of the making of the Calendar :

> The first idea that served as our basis is to consecrate, through the calendar, the agricultural system, & to bring the nation back to it, by marking the epochs & the fractions of the year by intelligible or visible signs taken from agriculture & the rural economy. [...] The calendar being something to which we so often resort, it is necessary to take advantage of the frequency of this use, to slip among the people the elementary rural notions, to show them the riches of nature, to make them love the fields, & methodically designate the order of influences of the sky & of the earth's produce. (Fabre d'Églantine, 1793, p. 6)
>
> The priests had assigned each day of the year the commemoration of a supposed saint: this catalogue presented neither utility nor method; it was the repertoire of lies, deceit or charlatanism. We thought that the nation, after having chased this crowd of canonised from its calendar, should find in its place all the objects that make up the true national wealth, the worthy objects, if not of its worship, at least of its cultivation; the useful produce of the earth, the tools we use to cultivate it, & the domestic animals, our faithful servants in these works; animals far more precious, undoubtedly, to the eyes of reason, than the beatified skeletons drawn from the catacombs of Rome. Consequently, we have arranged in order in the column of each month, the names of the true treasures of the rural economy. The grains, the pastures, the trees, the roots, the flowers, the fruits, the plants, are arranged in the calendar, in such a way that the place & the date that each production occupies is precisely the time & the day when nature presents it to us. (Fabre d'Églantine, 1793, p. 12)

The New Division of the Day

After the adoption of the decree instituting the new calendar and the new division of the day, Lagrange, a member of the Commission who supported the project, affirms very optimistically:

> The disadvantages that the decimal division of the day seems to present are more apparent than real, and on the contrary this division combines all the advantages that one can desire, both for the needs of life and for astronomical calculations. (Froeschlé, 1996, p. 310)

The workings of all clocks having to be changed, a decree of 9 February 1794 organises a competition for scholars and clockmakers to bring practical solutions.

> Decree of the National Convention of 21 st. Day of Pluviôse, second year of the French Republic, one and indivisible, which establishes a competition on the means of organising Watches and Clocks in decimal division.
>
> The National Convention, after having heard its committees of public instruction and finance, decrees:
>
> Article one. The artists of the Republic are called to compete for the solution of the following questions:

> 1st. Question. What is the simplest, the most solid, the least expensive organisation to give to pocket watches, to clocks, to measure, together or separately, the different parts of the day, which, by the decree of 4 Frimaire, is divided into ten hours, each hour into tenths, hundredths, thousandths and ten thousandths; and what is the best way to indicate them, either by the dial, or by the ring? (Decree of 21 Pluviôse year II, 9 February 1794)

The clockmakers then have little choice, in the midst of Terror, to refuse these proposals. Some pretty watches and a few rare clocks make their appearance, but without great success, most of the time with a double graduation, duodecimal and decimal.

However, this new division of the day is not well accepted by the population. On the one hand, humans do not easily adapt to changes like that of the measurement of duration, and on the other hand the division of the day into 10 does not lend itself well to division by four, for example. The divisors of 10 are fewer than those of 24. Moreover, there are at least two other reasons, linked to each other, and of an economic order, which can explain the failure of this reform. There is first the significant cost of the modifications that would have had to be made to the clock mechanisms of the time and, on the other hand, the lack of economic interest in making these modifications; indeed work was then paid by the task and not by the duration as now. The watches and clocks made according to the new standards remained few in number. One can discover a set of magnificent clocks made at this time at the Carnavalet museum in Paris.

Scientific and Pedagogical Aspects

The Problem

Our calendar and our division of the day, which seem natural yet complex to pupils, can be questioned. This then provides an opportunity for them to learn and reflect on this history of time and its organisation. The role of natural constraints and cultural aspects can thus be clarified in their eyes. The role that mathematics has played in this construction should convince young minds of their usefulness in human life. To motivate this questioning, comparison is often a good means of highlighting the invariants linked to nature, the different choices made by human societies.

The difficulty that the French population of the time encountered with the new decimal division of the day was even greater than that which they had to face with the new calendar. These two reforms both changed deeply ingrained habits. Yet one disappeared much faster than the other. The essential factor is certainly that one requires solving difficult conversion problems to obtain information as important as a duration or a schedule. It is well known to mathematics teachers that proportionality, which is a common thread in compulsory schooling, is not an easy concept and requires practice to master satisfactorily. The level of education of the population at that time was not what it is today. Does this mean that the adult population, in a somewhat general way, would be more comfortable with such conversions, even with a pencil in hand? Nothing is less certain. One only has to remember the apprehension of the French when switching to the euro. One training approach could be, counter-intuitively, to work on proportionality, the decimal system and fractions in the same concrete situation and thus adding complexity. Complexity is not necessarily synonymous with insurmountable difficulties if this complexity is immersed in a sufficiently familiar context, involving quantities that speak to pupils; it can, on the contrary, be a source of stimulation. If the concrete situation belongs to human history, it should arouse even more curiosity.

Pedagogical Aspects

Two activities, one on the Republican calendar and the other on decimal time, were proposed to classes of 10- and 11-year-old pupils. One involves gathering information from a complex document and should open the discussion on the mathematical, but also astronomical and sociological, aspects of a calendar. With the calendar, there is an opportunity to talk about the Earth-Moon-Sun system, the duration of revolution and rotation of the Earth, the role of mathematicians and astronomers in establishing a calendar, the social and political constraints that intervene and the gradual implementation of the decimal system. The second activity involves decimal, fractional and proportional calculations. It is more difficult, especially for 10-year-old pupils. With the correspondences between the decimal division and the traditional division of the day, the links between the different representations of numbers can appear more clearly, proportionality can also be translated in several frameworks, graphical, numerical... The proposed work goes beyond the knowledge required at this level of duration as a quantity, but, on the other hand, fits perfectly into the technical work expected on both fractional writing and calculation and the decimal system and calculation (Kim & Kang, 2014). A discussion on the historical aspects of this decimal division of the day, but also more generally on the implementation of the decimal system in society with its advantages and its sticking points, is an opportunity for pupils to discover the role that mathematics can play in everyday life.

Implementation in the Classroom

Several organisations have been set up for the running of these two activities. None exceed three sessions.

The Revolutionary Calendar A sheet, composed of a large Republican calendar accompanied by information and questions, is distributed to the pupils. After they have become aware of it, the teacher questions the pupils about the proposed situation, answers the questions that inevitably arise about the historical period, orally quizzes on vocabulary, asks for the rephrasing of questions, then lets them work, either individually or in pairs. The answers given to the questions are recorded on a copy that can either be handed in at the end of the activity, or later, in the form of homework when it comes to older pupils.

Internet research can be organised in various ways. Thus teachers can set up two pupils in front of the teacher's computer, while others may choose to work collectively with a projector.

Pupils are encouraged to debate with the teacher to question further about this calendar. Additional information of a historical, astronomical and social nature can then be provided. Generally, one session is enough to carry out this activity.

Decimal Time After a brief presentation of the historical context, the sheet on the decimal division of the day is distributed to the pupils. After a loud reading of the statement, an explanation and rephrasing phase are often necessary to properly set the situation. Questions arise about the historical context, to which it is necessary to take the time to answer to complete the introduction. The classroom procedure can then take different forms depending on the reactivity of the classes and the temperament of the teacher. You can let the pupils work in groups and independently, then make a synthesis of the answers, or propose a common reflection and the activity then takes the form of a debate. You can ask to write a report of the answers that have been developed either in class at primary school, or as homework at secondary school. Depending on the organisation adopted, the length of the debates and the explanations provided, two sessions are generally necessary to complete this activity.

The Reports

The Activity on the Republican Calendar (Appendix 2)

Sandrine (school in La Rochelle, Charente-Maritime) had her 10-year-old pupils do the activity on the Republican calendar. It took place over an hour and a half. The pupils worked in pairs. The main difficulty came from the poor readability of the calendar. The pupils had not noticed that the Republican calendar had references to the Gregorian calendar. After giving them this information, the teacher encouraged them to read all the small print. The pupils then worked independently.

1. Question 1: Half of the pairs stopped at reading the title and therefore answered YES.
2. Question 2: All the answers were correct.
3. Question 3: All pairs indicated 360 days without taking into account the sans-culottes' holidays.
4. Question 4: Proposed answers:

   - months divided into two columns, Gregorian and Republican,

- in three parts: the numbers in the first column, words in the second and the French calendar in the third,
- they are divided into 4, because $4 \times 3 = 12$,
- no answer,
- the correct answer.

Most have a horizontal reading of the document instead of vertical. No justification was given to the question.

5. Question 5: One group indicated the two names Romme and Fabre d'Églantine, five groups Romme only and two groups mistakenly give Philibert-Louis Delacourt.

   The 11-year-old pupils had much less difficulty in interpreting the document. There is an obvious evolution from one year to the next in the way pupils analyse the organisation of a document. Thus, most 11-year-old pupils knew that a year consists of 365 or 366 days and diligently searched at the bottom of the Republican calendar for the additional five or six days (see Inès' copy in Appendix 4).

The Activity on Decimal Time (Appendix 3)

To illustrate how a teacher can experiment with the activity on decimal time, here is an example of an analysis of decimal time carried out by Audrey with a class of 26 pupils aged 9 and 10 years old.

Presentation of the Sessions

Session 1

- *Reactivation*: decimal fractions with references to the display and manipulation on the number line.

  Introduction of the class clock and its operation (divided into 12, but 24 hours).
- *Towards interdisciplinary*: highlighting the changes of the French Revolution.
- *Presentation of decimal time*: manipulation, research and answers to questions 1, 2, 3 and 4.
- *Institutionalisation*: the pupils discovered decimal time and manipulated it. They guessed the answers to questions 2, 3 and 4, but could not justify them, as they arrive at the following equality: $10H = 24h$ (with help), which they cannot solve (Fig. 6.1).

The teacher then intervenes to explain that it is necessary to divide by 10.
The pupils arrive at $1H = 2.4h$ (Fig. 6.2).
Three pupils managed to convert $2.4h$ into $2h24min$.

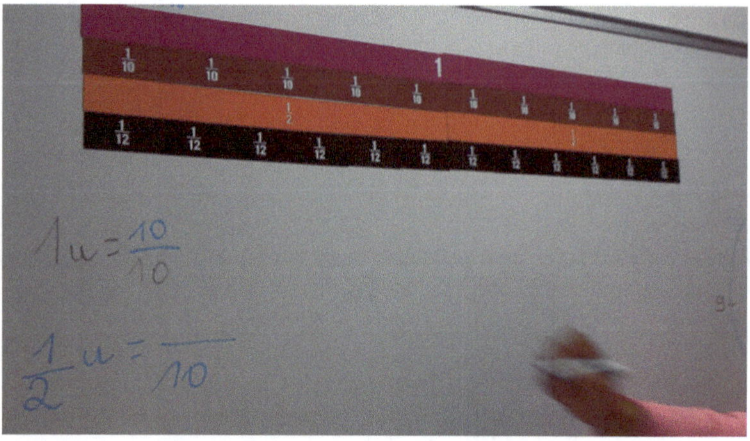

Fig. 6.1 The linear representation of the day (Photo : Audrey Fayollat)

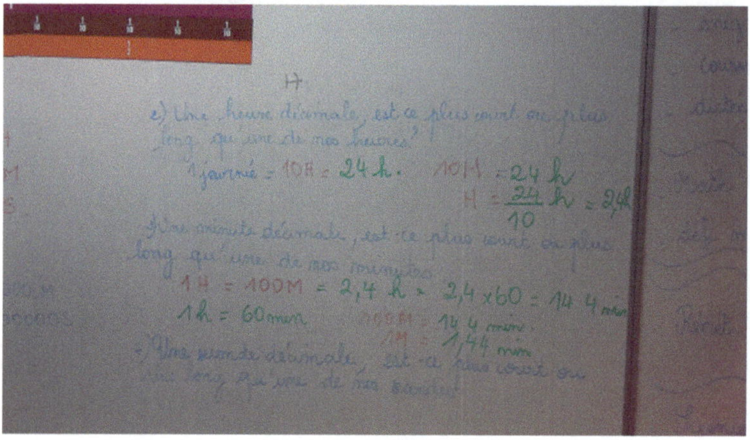

Fig. 6.2 The treatment of proportionality (Photo : Audrey Fayollat)

Session 2

- *Reactivation of session 1*: presentation of standard time/decimal time clocks (Fig. 6.3). Review of questions 1, 2, 3 and 4. The answers are correct, the pupils are very invested in the project.
- *Research phase around question 5*. How much does $1H$ (decimal time) represent in $h - min - s$?

  Review of the equality $1H = 2.4h$. Only three pupils managed to convert $1H$ into $2h24min$. Most proposed $2h$ and $40min$. A reminder was made about the conversion of a decimal notation into $h - min - s$.
- *Research phase around question 7*. What time does $3.40H$ represent for us? Before starting the research, the pupils observe the clock indicating $9h34min =$

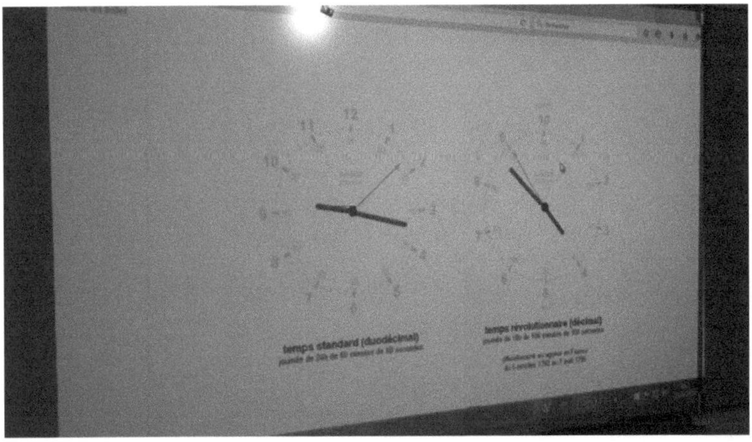

Fig. 6.3 The two clocks of the site http://www.procrastin.fr/blog/index.php?2005/12/05/61-horloge-decimale (Photo : Audrey Fayollat)

$3H98MIN$. On the written traces, all the procedures (sometimes erroneous) to be considered by the pupils for the resolution of a situation of proportionality are found:

- linearity (subtract 58 from $9h34$, but the pupil did not succeed)
- multiplication by 2.4: $3H40MIN \times 2.4$. This calculation was successful, as the pupils made the connection:
  $3H40MIN = 3.4H$ and were able to calculate 3.4×2.4. They did not convert $8.16H$ into $h - min - s$.

- *Structuring phase.* Question 6 was not addressed to avoid complicating the work. An expert strategy was highlighted to answer question 7: multiply the decimal time by 2.4 and convert the result into $h - min - s$.
- *Formative assessment.* Small transformations on slate were proposed:

$$5H = \cdots h;$$
$$2.5H = \cdots H \cdots MIN = \cdots h;$$
$$24h = \cdots H;$$
$$7.5H = \cdots H \cdots MIN = \cdots h.$$

- *Conclusion of the sessions on decimal time.* The enthusiasm of the pupils and the interest were such that an article was written on the school's website, which allows you to go to the link of the clocks and which offers some riddles (Fig. 6.4).

Prerequisites

The prerequisites needed for session 1:

- mastery of decimal fractions,

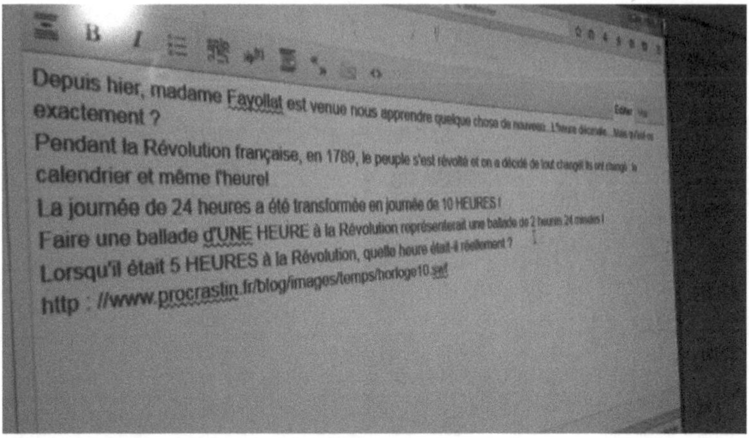

Fig. 6.4 A trace of the activity (Photo : Audrey Fayollat)

- mastery of the notions of durations and their relations.

The pupils can thus easily answer questions 1, 2, 3 and 4. For the justification of the longest, shortest and calculations (from question 5), these skills are not sufficient. They arrived at the equality $1day = 10H = 24h$ without problem, but could not arrive at the equality $1H = 24/10h = 2.4h$. However, by accompanying them in the calculations, they found the ratios between MIN and min and between S and s.

The prerequisites needed for session 2:

- the transformation of the hour into decimal notation in $h - min - s$.

The learning objective of session 2 is the resolution of a proportionality problem.

Points of Vigilance

- Particular attention will be given to the symbols (H, MIN, S and h, min, s). The use of different colours can be practical.
- The day marker, which is not modified, is an important marker, which can also be highlighted in colour.
- In the formulation of the questions, a reference to the pupil's experience can help their understanding: "I am going for a walk for an hour. Will this walk be longer now or in 1794?".

Some arrangements of the information resisted their sagacity, like the column containing the name of the days, and the parallel presentation in the document of the two calendars was not often perceived.

The pupils were interested in this activity, particularly by the motivations that led to the introduction of this change and by the reasons for its abandonment. Some expressed surprise and were destabilised, wondering why we were doing history in mathematics.

We can compare two characteristic copies, placed in appendices 3 and 4, on the treatment of this activity of decimal time according to the age of the pupils.

General Assessment

The activity on calendars is generally successful by the whole class, both for 10-year-old pupils and those aged 11, even if they were surprised by this mix of mathematics and history. The natural curiosity of some pupils leads to very rich exchanges that bring a lot of information to the whole class. This allows, in an accessible and obvious way, to illustrate a contribution of mathematics to human activity.

The activity on the decimal division of the day was better achieved by 11-year-old pupils, which is normal, since the decimal system is still being constructed for many 10-year-old pupils. At this level, it was necessary to guide them more. The proportionality situation was not recognised by the vast majority of pupils, the teacher's help was almost always necessary at this stage. Different treatments of this proportionality were then implemented. Surprisingly, while the difficulties with conversions are real, the context and the new duration magnitude motivated the pupils who, for the majority of them, were active in their research, as evidenced by the various session reports provided by the teachers. The anchoring of the activity on the duration magnitude seems to have allowed them to better appropriate the questions asked, to better model, to better control the proposed results. Comment from a teacher: "The difficulty motivated a good half of the class who searched with enthusiasm (and these were not necessarily the best pupils)."

Most of the teachers who proposed these activities needed two sessions, sometimes three. It was not too time-consuming. The feedback from the pupils is encouraging. They appreciated, in their majority, the debates and the contributions of knowledge where they could be active and in a situation of research.

Ways to Go Further

To extend the proposed activity on the calendar, it can be interesting to also work on our Gregorian calendar. This will be more of an arithmetic work with the use of Euclidean division (number of weeks, day shift from one year to the next, drift of the Julian calendar, date of Easter day...). The study of a lunar calendar can also be considered. To continue on the theme of decimal time, we can, of course, work on the technique by setting up a series of decimal hour-traditional hour conversions to automate calculations. Or even make them think about a problem situation like this one:

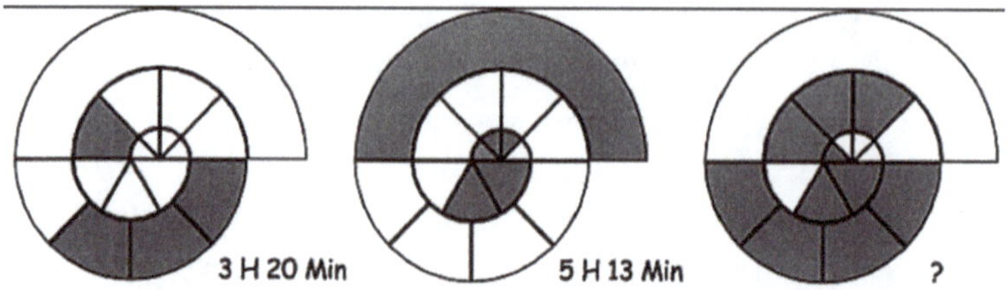

Fig. 6.5 The three shells of Jacquot (Rallye Mathématique de Poitou-Charentes in 2015), source: https://www.apmep.fr/IMG/pdf/Epreuve_2nde_pro.pdf

Fig. 6.6 The Berlin Clock: the *Mengenlehreuhr* displaying 10:31

Citizen Jacquot the Snail has switched to the decimal time of the Revolution. His shell marks all the decimal hours of the day to the decimal minute. Each minute, one or more portions of crowns colour or become clear again. Here is what his shell looks like at different times of the day (Fig. 6.5). If you have understood the logic of this display, give the decimal time displayed by the third shell.

To stay in the spirit of this last example, we can propose to understand other possible presentations of our time display. It could thus be interesting to study the reading mode of the astonishing clock installed in Berlin (Fig. 6.6) and why not imagine others. It's still arithmetic!

Acknowledgments This chapter was written by Frédéric de Ligt, IREM&S de Poitiers (France). Thanks to the teachers who took part in this experiment: Audrey Fayollat (Jacques-Yves-Cousteau primary school, Vouneuil-sous-Biard), Cyril Redondo, Matthieu Gaud and Sandrine Bonnel-Ventura (Pierre-Mendès-France secondary school, La Rochelle).

Appendix 1: Concepts and Skills Addressed in This Chapter

Concepts Covered

- Use and represent sexagesimal fractions, decimals and decimal numbers.
- Solve problems that simultaneously mobilise different units of time measurement and conversions.
- Identify a situation of proportionality between two measures of duration.
- Master the units of time measurement and their relationships.
- Extract numerical data from a calendar.

Skills Developed

- **Search and research skills**: Organise information extracted from a calendar. Make hypotheses.
- **Reasoning**: Interpret the meaning of a calculation, a numerical result, an information.
- **Computational skills**: Switch from decimal notation to fractional notation and vice versa.
- **Mathematical communication skills**: Debate with arguments, write a report presentable at the end of the research.
- **Mathematical representation ability**: Use diagrams, drawings to represent a duration problem.

Appendix 2: The Republican Calendar. A Bit of History

During the French Revolution, a new calendar was instituted in France by decree of the National Convention on November 24, 1793.

Equality Liberty

Decree of the National Convention on the era, the beginning and organisation of the year, and the names of the days and months

From the 4th day of Frimaire, the second year of the French Republic, one and indivisible

The National Convention, having heard its committee of public instruction, decrees the following:

Article One.

The era of the French counts from the foundation of the Republic, which took place on 22 September 1792 of the vulgar era, the day when the sun reached the true autumn equinox, entering the sign of Libra at 9 hours 18 minutes 30 seconds in the morning, for the Paris observatory.

II. The vulgar era is abolished from civil uses.

III. Each year begins at midnight, with the day when the true autumn equinox falls for the Paris observatory.

IV. The first year of the French Republic began at midnight on 22 September 1792, and ended at midnight, separating 21 from 22 September 1793.

This calendar was used until 1806. After that, France returned to using its old Gregorian calendar.

A Few Questions

Look at the Republican calendar available on https://clespourlhistoire.ac-besancon.fr/le-calendrier-revolutionnaire-et-ses-symboles/.

1. Can we say that this is the calendar for 1793?
2. How many months are there in a Republican year, and how many days in each month?
3. How many days are there in a Republican year?
4. In how many parts is each month divided? Why do you think this is?
5. Search the Internet for information on the person who chaired the commission responsible for drawing up this calendar, as well as the person who found all the names on the calendar.

Note:

As in our calendar, it was intended to make this civil calendar coincide with celestial movements by adding a day to the year every four years, on the day of the Revolution, the year was then called "sextile".

Every four years—the duration of a *Franciade*—a sixth additional day was added, called "Revolution Day".

Appendix 3: The Revolutionary Hour. A Bit More History

Article XI of the same decree also defines a new decimal time:

> XI. The day, from midnight to midnight, is divided into ten parts or hours, each part into ten others, and so on to the smallest measurable portion of duration. The hundredth part of the hour is called a decimal minute; the hundredth part of the minute is called a decimal second. This article will be strictly for public acts, from 1st Vendémiaire, the third year of the Republic.

The day begins at midnight, so at noon it was 5 o'clock, but 5 hours of the new hour, and at midnight it was 10 o'clock, 10 hours of the new hour.

"Decimal" watches were built at the time, but decimal time was not used for long, being abolished in 1795.

A Few Questions

The symbols h, min, s stand for hours, minutes and seconds. We will capitalize H, MIN and S for the decimal hours, minutes and seconds at the time of the French Revolution.

1. How many decimal minutes are there in a day? How many decimal seconds?
2. Is a decimal hour shorter or longer than one of our hours?
3. Is a decimal minute shorter or longer than one of our minutes?
4. Is a decimal second shorter or longer than one of our seconds?
5. In $h - min - s$, how much is a decimal hour H?
6. The time is 3pm. What time would it be in decimal Revolutionary time?
7. It's $3.40H$ on the old decimal watch above. Exactly what time does this represent for us?
8. Suggest one or more reasons why this new division of the day might have failed so quickly?

Note:

A comment posted on a site that presents the animation of a decimal watch: http://www.procrastin.fr/blog/images/temps/horloge10.swf

> With standard time, there are 86,400 seconds in a day, compared to 100,000 with decimal time. Seconds pass faster in decimal time. So, before tinkering with my clock, I expected (stupidly) to have a second hand that spins faster than on our watches. But look at the flash animation. The opposite happens: the decimal second hand moves slower!!! Well, when you think about it, it makes sense: OK the seconds are shorter, but there are more in a round of the dial. But you still have the strange feeling of a slower time. In any case, it has that effect on me. Not you?

Appendix 4: The Copy of Inès, a 11-Year-Old Pupil

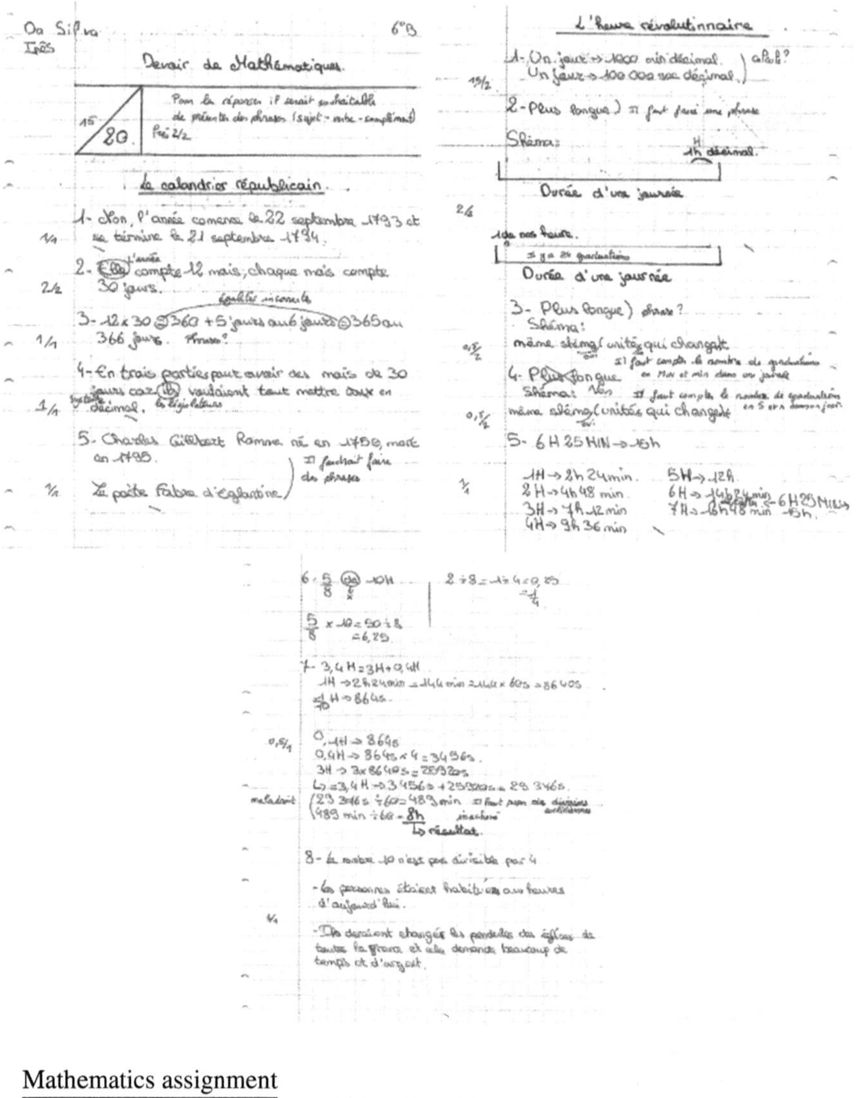

<u>Mathematics assignment</u>

15/20. For the answers, it would be advisable to present sentences (subject, verb, complement)

<u>The Republican calendar</u>

1. No, the year begins on 22 September 1793 and ends on 21 September 1794.
2. It *(the year)* has 12 months, each year has 30 days.

3. $12 \times 30 = 360 + 5$ days or 6 days $= 365$ or 366 days *(inconsistent equality) (sentence?)*
4. In three parts to have months of 30 days because they *(the legislators)* wanted to put everything in decimal *(system)*.
5. Charles Gilbert Romme born in 1756, died in 1795 *(a sentence should be made)*
6. The poet Fabre d'Églantine *(a sentence should be made)*

The revolutionary hour

1. One day > 1000 decimal min
   One day > 100,000 decimal sec *(calculations ?)*
2. Longer *(a sentence should be made)*
   Diagram: duration of a day represented by a segment divided into ten equal parts, each part representing 1 decimal hour, the same segment representing the duration of a day is divided into 24 equal parts, each part corresponding to 1 of our hours. These parts are smaller.
3. Longer *(sentences?)*. Same type of diagram *(yes)*. The units change *(you have to count the number of graduations in MIN and min in a day)*
4. Longer *(no)*. Same pattern with changing units *(yes) (You have to count the number of graduations in S and s in a day)*
5. (actually 6)
   $6H25MIN > 15h$
   $1H > 2h24min$
   $2H > 4h48min$
   $3H > 7h12min$
   $4H > 9h36min$
   $5H > 12h$
   $6H > 14h24min$
   $6H25MIN > 15h$ *(to detail)*
   $7H > 16h48min$
6. $\frac{5}{8}$ of $10H$
   $\frac{5}{8} \times 10 = 50 : 8 = 6.25; 2 : 8 = 1 : 4 = 0.25 = \frac{1}{4}$
7. $3.4H = 3H + 0.4H$
   $1H > 2h24min = 144min = 144 \times 60s = 8640s$
   $\frac{1}{10}H > 864s$
   $0.1H > 864s$
   $0.4H > 864s \times 4 = 3456s$
   $3H > 3 \times 8640s = 25,920s$
   $3.4H > 3456s + 25,920s = 29,376s$
   $29,376s : 60 = 489min$
   $489min : 60 = 8h$ *(you have to do Euclidean divisions) (unfinished)(awkward)*
8. - The number 10 is not divisible by 4.
   - People were used to today's hours.
   - They had to change the clocks of all the churches in France and this requires a lot of time and money.

Appendix 5: The Copy of Chloé, a 10-Year-Old Pupil

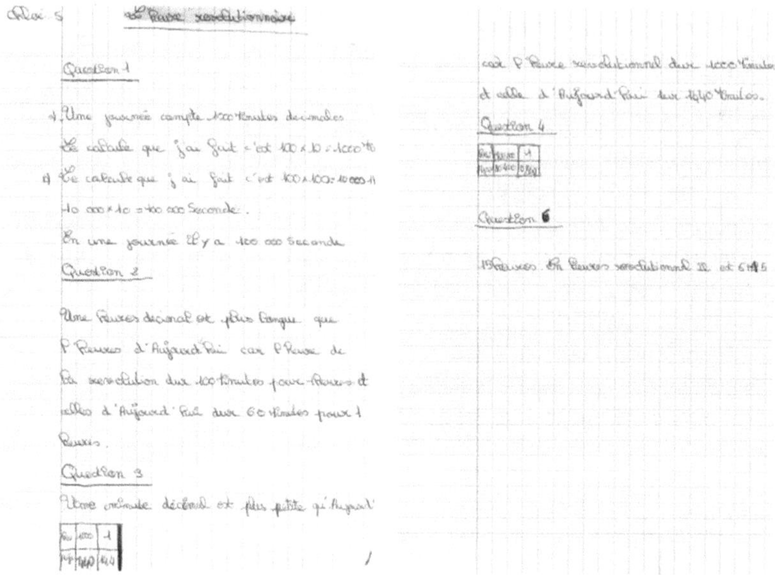

Revolutionary time

1. <u>Question 1</u>

   a. A day counts 1000 decimal minutes.
      The calculation I made is $100 \times 10 = 1000 MIN$

   b. The calculation I made is $100 \times 100 = 10{,}000 S$ in $1 H$;
      $10{,}000 \times 10 = 100{,}000$ seconds.
      In one day there are 100,000 seconds.

2. <u>Question 2</u>

   A decimal hour is longer than today's hour because the French Revolution hour lasts 100 Minutes for 1 hour and today's lasts 60 minutes for 1 hour.

3. <u>Question 3</u>

   A decimal minute is smaller than today's because the revolutionary hour lasts 1000 Minutes and today's lasts 1440 Minutes.

Revolution	1000	1
Today	1440	14.4

4. Question 4

Revolution	100,000	1
Today	86,400	0.864

5. Question 6
   15 hours, in revolutionary time it is 6:15

References

Assemblée Nationale Française. (1791). Décret de l'Assemblée Nationale du 30 mars 1791. http://mjp.univ-perp.fr/france/1793mesures.htm

Assemblée Nationale Française. (1793). Décret du 1er août 1793. http://mjp.univ-perp.fr/france/1793mesures.htm

Assemblée Nationale Française. (1795). Adoption du système métrique décimal (18 Germinal an III, 7 avril 1795). https://mjp.univ-perp.fr/france/1793mesures.htm

Convention Nationale Française. (1792). Décret du 20 décembre 1792. https://books.google.fr/books?id=9Uk2iVhY-bsC&printsec=frontcover&hl=fr#v=onepage&q&f=false

Convention Nationale Française. (1793a). Décret de la Convention Nationale concernant l'ère des Français (5 octobre 1793). https://mjp.univ-perp.fr/france/1793ererep.htm#1

Convention Nationale Française. (1793b). Décret de la Convention Nationale du 4e jour de Frimaire de l'an second de la République française. https://archive.org/details/decretdelaconven00fran_41

D'Alembert, J. l. R. (1754). Article "décimal". In *Encyclopédie, ou Dictionnaire raisonné des sciences, des arts et des métiers* (Vol. 4, pp. 668b–670b). Paris. https://enccre.academie-sciences.fr/encyclopedie/page/v4-p684/

Dowson, J. (1891). Decimal coinage, weight, and measures. *Journal of the Society of Arts*, *39*, 201–226.

Fabre d'Églantine, P.-F.-N. (1793). Rapport fait à la Convention Nationale dans la séance du 3 du second mois de la seconde année de la République française, au nom de la Commission chargée de la confection du calendrier. https://gallica.bnf.fr/ark:/12148/bpt6k48746z

Froeschlé, M. (1996). À propos du calendrier républicain : Romme et l'astronomie. *Annales historiques de la Révolution française*, *304*(1), 303–325.

Kim, E., & Kang, H. (2014). A construction of decimal fraction of elementary mathematics textbook and analysis of students state of understanding based on measurement theory. *Journal of Elementary Mathematics Education in Korea*, *18*, 37–62.

Michelet, J. (1853). *Histoire de la Révolution française* (Vol. 7). Le Vasseur.

Perovic, S. (2012). *The calendar in revolutionary France: Perceptions of time, in literature, culture, politics*. Cambridge University Press.

Romme, C.-G. (1793). Rapport sur l'ère de la République. https://ia801604.us.archive.org/1/items/rapport00unse_1ih/rapport00unse_1ih.pdf

Thiers, A. (1866). *Histoire de la Révolution française* (Vol. 1). Lheureux.

Vera, H. (2008). Decimal time: Misadventures of a revolutionary idea 1793. *Journal for the Study of Time, Kronoscope*, *9*(1), 29–48.

Zupko, R. (1990). *Revolution in measurement, western European weights and measures since the age of science*. American Philosophical Society. Memoir 186

Frédéric de Ligt is a mathematics teacher at the Barbezieux high school. A member of the history of mathematics group at the IREM&S in Poitiers, he experiments with students by introducing the history of mathematics into the curriculum. Deeply involved in all aspects of mathematics education, he is the president for the Poitou-Charentes region of the APMEP (Association of Mathematics Teachers in Public Education) and writes the problems column for the APMEP's journal, *Au fil des maths* [Along the Lines of Mathematics].

Chapter 7
What If We Measured the Schoolyard? Surveying Experiences

Marc Troudet

> *It will be necessary to practice a lot to determine regular areas ; that of a door, a window, a table, the floor or the ceiling of a room, a courtyard, etc.* [...] *one will become familiar in this way with the simplest applications of the metric system, one will get the intuition, one will have the correct notion of the use of the various units; and this acquisition* [...] *will later be of considerable help, during the period of studies.*
>
> Charles-Ange Laisant (1906, p. 79)

Abstract After providing the historical, social, and professional reasons that led to the resolution of practical geometry problems in primary schools in France during the Third Republic (1870–1940), this chapter describes surveying experiments carried out at school and in the early stages of secondary school. It shows how the use of a surveyor's square and measuring instruments allows for the measurement of a school's courtyard and the benefits that pupils gain from these practical exercises in learning the concept of area.

Keywords Practical geometry · Surveying · Area · Manipulation · Instruments · Measurement · Courtyard

In his *Cours complet d'arpentage élémentaire, théorique et pratique* published in 1852, David Puille, professor of mathematics and physics in Paris, gives the following definition of the verb "to survey" :

> To survey an agrarian surface is to seek how many times this area contains another taken as a unit: the result expresses the extent of the surface being measured. (Puille, 1887, p. 163)

M. Troudet (✉)
IREMI de Grenoble, Saint Martin d'Hères, France
e-mail: marc.troudet@univ-grenoble-alpes.fr

M. Moyon et al. (eds.), *Teaching Mathematics Through Historically-Based Activities*, History of Mathematics Education,
https://doi.org/10.1007/978-3-032-16990-7_7

The author then specifies that the original meaning of the verb to survey is "to bring surfaces back to the arpent", an old unit of measurement used in France. Surveying is thus "the art of measuring the area of a piece of land" (F.I.C., 1882, p. 1), therefore an "application of geometric principles on the measurement of areas" (Buisson, 1887, pp. 120–121). In France, the Third Republic, from 1870 to 1940, is marked by the generalisation of primary education. In mathematics, the topics of areas and volumes occupy an important place. The teaching of these quantities is then "focused on practical problems" (Perrin, 2005, p. 306) such as surveying and calculating volumes of everyday objects.

During the 1960s and 1970s, the so-called modern mathematics period saw the place of these concrete questions reduced in favour of numbers. At the same time, social practices evolved and the instruments once used for these operations gradually disappeared. Since 2008, the teaching of quantities has made its return to the programs while primary and secondary school teachers have rarely received training on their usual applications. Concurrently, from the 1980s to the present day, large-scale studies on the assessments of pupils aged 11 to 15 have been conducted in France (Moreira Baltar, 1997, p. 47), the United States, England and Australia (Lehmann, 2003, p. 531). In learning the concept of area, they highlight several difficulties: confusion between perimeter and area, unit errors and lack of knowledge of area formulas and errors. Many pupils also struggle to calculate the area of composite surfaces, even when they are superimposed on square grids.

In this chapter, we begin by detailing the reasons that led to the use of practical problems in the teaching of areas in primary school during the Third Republic. We take the opportunity to recall that the teacher was often the village surveyor, which is largely unknown today. This part can thus serve as a basis for a historical presentation in class.

We then detail the different stages of a survey, its instruments and methods. We then examine the difficulties related to the teaching of areas. We show at this stage how to differentiate over three years a problem situation: the surveying of a schoolyard. This journey seems to us capable of remedying the main difficulties related to the teaching of the concept of area. Finally, we describe the resolution of this problem, dealt with alternately between the classroom and the field. The surveying of the courtyard is part of a progression of learning the concept of area over three years for pupils aged 10 to 12. Calculating an area using a grid at 10 years old, then using formulas to calculate areas of increasingly complex figures between 11 and 12 years old. As suggested by Charles-Ange Laisant (1906, p. 79), these exercises emphasise the manipulation of measuring instruments and the use of concrete supports from the pupil's environment. They lead him to build empirical benchmarks on areas. Carried out in the field or in class and spread over a year, the stages of measurement mobilise multiple tools: measuring instruments and ICT (Information and Communication Technologies for Teaching) including the calculator. We explain how this complex task can be differentiated over three years in order to allow each pupil to estimate the area of the courtyard he "surveys" every day.

Historical Context

Surveying in the Primary Education Program

The introduction of surveying teaching in primary schools results from an evolution of the programs initiated under the *Ancien Régime*. Renaud d'Enfert explains that surveying lessons are then given by some schoolmasters. But, despite the support of Mirabeau or Condorcet, it was not until 1816 that teachers holding the first degree certificate (i.e. the most qualified, who usually work in major cities) saw surveying officially enter their duties (D'Enfert, 2003b, p. 42).

In 1833, Guizot, Minister of Public Instruction, judged that "the elements of practical geometry [...] provide the first data of all industrial professions" (D'Enfert, 2003b, p. 67). Indeed, this discipline uses the properties of geometry to measure lines, surfaces and solids, in order to solve everyday life problems. The Guizot law of 1833 inscribes primary education in geometry in a double tradition: that of measurement and practical geometry on the one hand, and that of drawing on the other hand (D'Enfert, 2003a, p. 42). The elements of geometry, hitherto reserved for secondary education, are taught from primary school. They constitute "the theoretical basis of drawing [...] and surveying" (D'Enfert, 2003b, p. 67).

As early as 1836, teachers are invited to conferences educational or advanced courses in which "the legal system of weights and measures and surveying occupy a significant place" (D'Enfert, 2003b, p. 78). They are even then introduced to a more precise method of surveying than those they teach in their classes.

The Surveying Schoolmaster

In fact, these trainings have another purpose. On the one hand, they contribute to the teaching of these concepts. On the other hand, they allow schoolmasters to develop their practice of surveying and to carry out surveying activities. This additional activity is therefore not only tolerated but encouraged. Indeed, it allows schoolmasters to "establish more contact with the population and thus contribute to the development of primary education" (D'Enfert, 2003b, p. 80). The primary school is then the only official instruction body for surveying and schoolmasters make up for the absence of surveying offices, especially in the countryside, a situation that persists beyond the Third Republic. From the beginning of the article dedicated to surveying in the *Nouveau dictionnaire de pédagogie et d'instruction primaire* (Buisson, 1911), the author Henri Sonnet, honorary academy inspector, recalls how the study of surveying in normal schools offers schoolmasters "a means of making themselves useful later in the municipalities, and thus creating a resource to add [to their] modest emoluments". Moreover, despite "the prohibition of exercising industrial or commercial professions" pronounced by article 32 of the law of 15 March 1850, one cannot

> forbid a schoolmaster engage in surveying operations, provided that this ancillary work always took place outside of class time and did not in any way interfere with their preparation. (Buisson, 1887, p. 120)

The schoolmaster is also encouraged to supplement his surveying knowledge with "the measurement of logs and the gauging of barrels" because

> the surveyor who is familiar with these geometric operations is often called upon to provide great services in the municipality he lives in, and anything that can contribute to increasing the legitimate influence of the schoolmaster is, in the end, beneficial to primary education. (Buisson, 1887, p. 121)

The Practice in Primary Education of the Third Republic

Primary education under the Third Republic will also be characterised by the "exercise of the senses" (D'Enfert, 2003b, p. 144) and this direction of education is already announced by the primary inspector Jean-Jaques Rapet in March 1855, in a pedagogical review close to the ministry. He argues that education should not be limited to "purely theoretical and abstract lessons" but should be accompanied by practical exercises "much more attractive to pupils". Regarding tools and examples of situations, he writes:

> Let us therefore train our pupils to measure all kinds of quantities, lengths, surfaces and volumes. In the absence of the instruments that the school does not own, the parents of our pupils, a merchant, a manufacturer, a builder, will be happy to lend us all the weighing and measuring instruments they own. If necessary, a metre alone, with a string or a thread ribbon, will be enough for us to get our pupils used to measuring almost all sizes, to find the surface of a field, a garden, a courtyard, a wall, to find the volume of a pile of earth, sand or manure, the quantity of materials that go into the construction of a wall, the number of metres of earth to extract or transport to do a certain job, the capacity of a vat, a ditch, a barn or a attic. (Rapet, 1855, p. 122)

The teaching programs include incentives for practice in addition to the traditional content. Those of the upper primary schools for boys (between 11 and 13 years old), published on 21 January 1893, place the study of surveying and the drawing up of plans in the second year. The paragraph begins with a calendar note and a call to leave the classroom:

> These last two courses will be held in summer and exclusively in the field, where the pupils will sketch during the operations. Back at school, they will have to report their plans to the indicated scale. (D'Enfert, 2003b, p. 262)

The training of pupil-teachers also prepares the future teacher to lead his pupils in the field. In the "geometry" paragraph of the *Special Instruction on the application of teaching programmes in primary normal schools* of 18 October 1881, the emphasis is on these applications. These

> occupy almost the entire third year, and three hours per week are allocated to them: this shows the importance that should be attached to the practical side of this science. (D'Enfert, 2003b, p. 206)

The teacher is encouraged to familiarise his pupils with the use of surveying instruments and to frequently exercise them on "flat and small area" terrains but also on "rough and quite extensive" terrains (D'Enfert, 2003b, p. 206).

These principles are thus recalled in the primary instruction manuals we have consulted. *L'Initiation mathématique* by the mathematician Laisant, is intended by its author as a "guide" from which "the educator should draw inspiration" (Laisant, 1906, p. 4). In this work, reissued 17 times, the frequent determination of areas of surfaces from the environment is a way of "getting the intuition" [of the metric system] and having "the correct notion of the use of the various units". Laisant believes that "this acquisition, useful in itself, will later be of considerable help during the study period" (Laisant, 1906, p. 79). Several primary instruction manuals consulted contain similar incentives. From the introduction of their *Manuel d'arpentage pour les écoles primaires*, the Brothers of Christian Instruction indicate that "serious study and blackboard demonstrations" of surveying methods are as important as the pupil's application of "the principles that have been explained to him" and handling "the instruments that he will only know perfectly after quite a long use" (F.I.C., 1882, p. vii). This warning concludes the preface: "It must be well understood that only practice can make a skilled surveyor" (F.I.C., 1882, p. viii). A few years later, the teachers Boucheny and Guérinet still write in a footnote of the theoretical principles of their *Géométrie au cours complémentaire* :

> We recommend as interesting and profitable exercises the surveying of the schoolyard, the square, a garden, a field. (Boucheny & Guérinet, 1912, p. 298)

Scientific and Pedagogical Aspects

In this part, we begin by describing the instruments, operations and methods of a survey taught in the primary schools of the Third Republic. We strictly followed these recommendations and used the same tools during the measurements carried out in the field with our classes.

We show how to articulate the measurement of a courtyard with a progression giving a large place to the manipulation of instruments and the measurement of surfaces from the pupil's environment. We believe this approach can build empirical benchmarks on area units and address the main difficulties related to teaching the concept of area.

We also revisit one of the objectives of surveying: developing vision in space.

The Instruments of Surveying

The equipment useful for a survey is limited to four instruments: "pegs, a surveyor's chain, cards and the surveyor's square" (F.I.C., 1882, p. 1). Henri Sonnet indicates

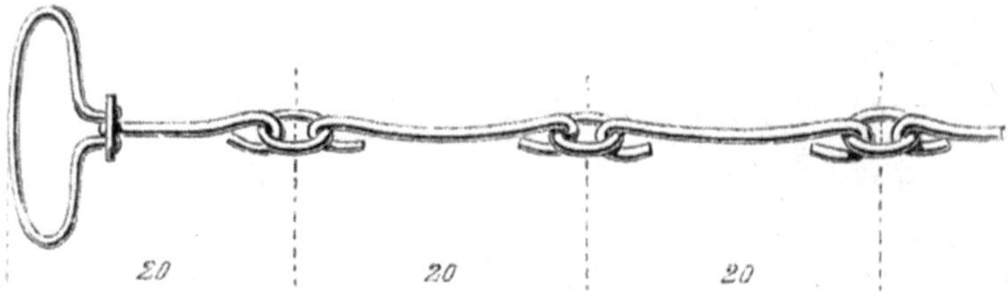

Fig. 7.1 Surveyor's instruments : Chain (F.I.C., 1882, p. 2)

Fig. 7.2 Surveyor's instruments : Tape measure wheel (Eysséric, 1855, p. 24)

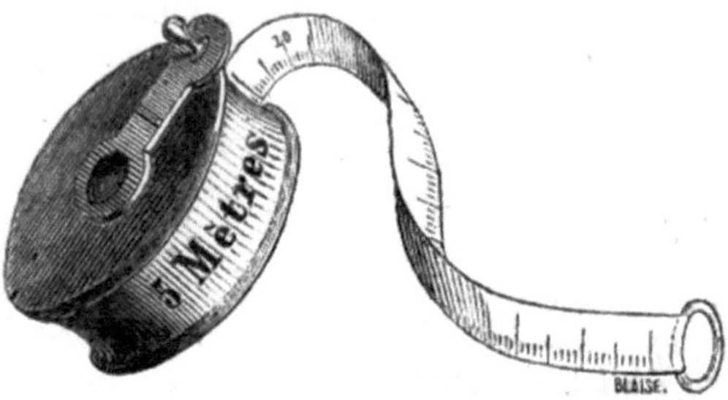

that lifting a surface can be done with the chain alone if it is "uncovered, substantially flat, and accessible in its interior" (Buisson, 1887, p. 120). However, since the survey is carried out to determine an area, it is more convenient to use the chain and the surveyor's square; this method applies to almost all small surfaces.

The pegs are wooden rods one and a half to two metres long and thirty centimetres thick. The lower end is finished with an iron point so that it can easily penetrate the ground. The difficulty in the field is to see the pegs, so the upper part is equipped with a small board, called a marker or simply a piece of white paper inserted in a slot made in the rod (F.I.C., 1882, p. 1). To operate with great accuracy and ensure the verticality of the rod, a plumb line can also be added.

The surveyor's chain (Fig. 7.1) is a wire chain ten metres long. It is made up of 50 links of 20 centimetres each, including the rings that join them (Hue & Vagnier, 1893, p. 312). Copper rings mark the metres and the middle of the chain carries a small iron rod. Surveyors also use the decametre tape, a steel tape ten metres long, wound on a wooden disc for easy transport.

However, the tape measure (Fig. 7.2) or pocket chain (F.I.C., 1882, p. 3) is preferred because its "flexibility allows it to follow the angular contours of the line" (Eysséric, 1855, p. 23). Made of a waterproof fabric tape ten metres long, wound around an axis and enclosed in a cylindrical box, this instrument is light and portable. Despite its superior convenience to the previous ones, it "offers little precision, as use and humidity significantly alter its length" (F.I.C., 1882, p. 3). It should therefore be reserved for "lengths of little surface and on which errors can only be very little prejudicial".

The surveyor's cross (Fig. 7.3) is the instrument that allows you to "draw lines perpendicular or parallel to others" and "determine the direction of an alignment"

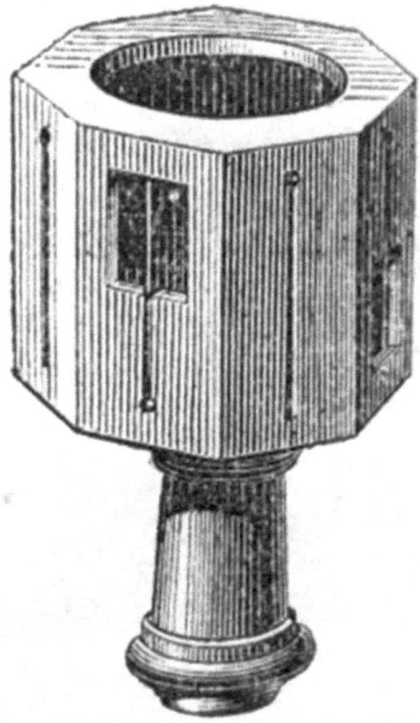

Fig. 7.3 Surveyor's instruments: Surveyor's square (F.I.C., 1881, p. 6)

(Puille, 1887, p. 118). This small copper box usually takes the form of a right prism whose base is a regular octagon. Four faces opposite two by two and at right angles, each have a slot and an opening called a window.

The slot of one of the faces corresponds to the window of the opposite face and vice versa. A horsehair or any other very fine thread is stretched vertically in the middle of each window, in line with the slot. The slot and window assembly is called a pinnule (F.I.C., 1882, p. 4,). The surveyor's square is perched on a foot, referred to as a "square foot or stick". A stem of one metre twenty to one metre forty in general, it is ironed at the end to drive it into the ground. A three-legged foot is used when the surveyor operates on rocky terrain. Appendix 2 describes instruments that can replace the surveyor's square if the teacher does not have one. A socket is screwed to the lower part of the square, in order to adapt the instrument on the square foot. Once the operation is done, the socket can be unscrewed and screwed inside the instrument to facilitate its transport (Puille, 1887, p. 119).

Field Surveying Operations

There are three surveying operations:

> to draw a straight line on the ground, to measure a portion of a straight line using the chain, to draw a perpendicular from a given point to a determined straight line on a terrain. (Puille, 1887, p. 165)

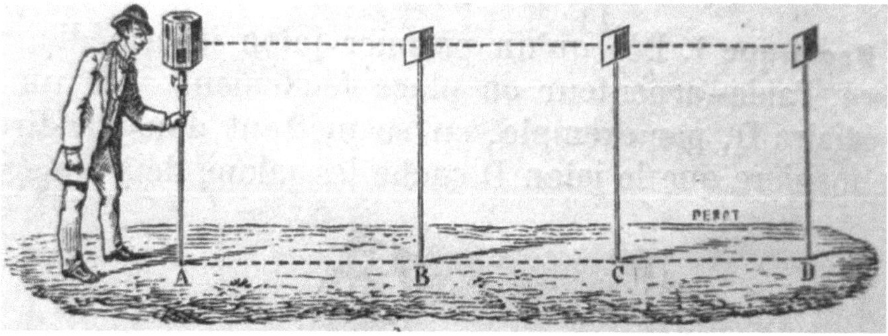

Fig. 7.4 Field surveying operations (1) (Frères des écoles chrétiennes, 1899, p. 8)

The "plotting" of straight lines on the ground can be done with a well-stretched cord if two points are not too far apart. However, if the distance between two points is significant, the straight line can no longer be drawn and we then content ourselves with marking several points at a suitable interval in its direction using stakes.

At least two people are necessary: the operator and his assistant. The assistant goes "ahead and is guided by the operator. It is the assistant who places the stakes at the points indicated by the operator" (Pécherat, 1897, pp. 28-29). Communication between the operator and his assistant is mainly done by gestures. In the description of this "exercise to be done in the schoolyard", Boucheny and Guérinet write:

> the operator signals with his hand, to his assistant, to place the peg more to the right or more to the left, until the peg is on the (correct) visual radius. Then he lowers his hand, and the assistant drives the pegs into the ground. (Boucheny & Guérinet, 1912, p. 292)

This operation, called staking a line or drawing an alignment (Fig. 7.4), requires a certain number of "trial and errors" (Puille, 1887, p. 122). It can be performed only with pegs but the use of a square for sighting makes it more precise.

The second operation is the measurement of a straight line AB (Fig. 7.5). The operator stands at one end and his assistant walks ten metres with the surveyor's chain. When the chain tightens, the assistant does not turn around but "the chainman shouts: to the right or to the left until the file placed vertically is in the direction AB". It is also "recommended for the assistant to be a little to the left of the line to be measured, so that his right hand is in alignment" (Hue & Vagnier, 1893, p. 316).

The surveyor must also know how to draw perpendiculars to lines staked out on the ground. Two situations are possible: either the point is on the line, or it is outside the line. In the first case, the surveyor must raise a perpendicular to a line AB from a point C located on the line AB. The square is placed at point C. The operator starts by making sure that point C is indeed on line AB. To this end, he successively sights the stakes A and B through the pinnules of the instrument. If he does not see them, he moves the surveyor's square until he sees them. An intermediate peg, placed on line AB, makes it easy to place the surveyor's square in alignment (F.I.C., 1882, p. 15).

Fig. 7.5 Field surveying operations (2) (F.I.C., 1882, p. 11)

Fig. 7.6 Field surveying operations (3) (Frères des écoles chrétiennes, no date, p. 153)

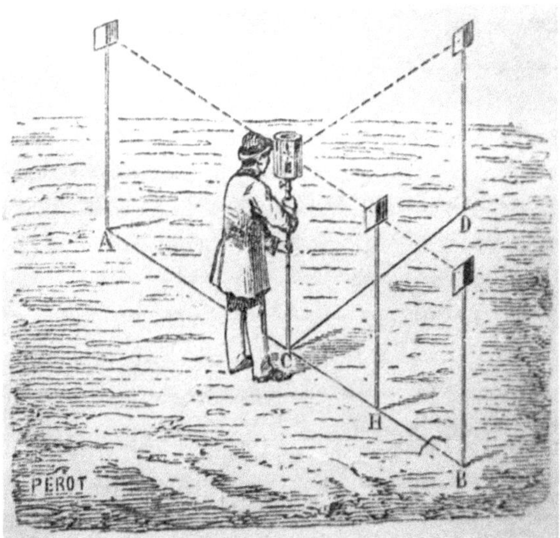

With the help of this instrument, the operator guides his assistant to place a peg D in a direction perpendicular to AB then he stakes out the line CD. In the second case, the surveyor must raise a perpendicular to a line AB from a point D located outside the line AB (Fig. 7.6). He positions himself on the line AB "and plants the square on this line, at the place he judges to be the foot of the perpendicular to be led" (Boucheny & Guérinet, 1912, p. 295). The operator then aims at the pegs A and B to ensure their alignment with the square. He then aims at the stake D and moves the square if necessary. David Puille adds that "it takes practice to determine the foot of such a perpendicular" and that a lot of practice is necessary to be able to "find it without much fumbling" (Puille, 1887, p. 123). The operator finishes by placing a peg at the point C obtained and then stakes out the line CD.

The Main Surveying Methods Taught in Primary School

Surveying follows the theoretical teaching of geometry given in class. The shape of the measured lands gradually becomes more and more complex: a triangle, a parallelogram, a trapezoid, convex polygons or with re-entrant angles. Sometimes, the plot is adjoined by a curved line, more or less sinuous like a river, it is then reduced to "a polygonal contour, considering the curve as a broken line" (Buisson, 1887, p. 121).

The measurement of the area of the land is organised in three stages. The surveyor begins by walking it (hence the other meaning of the verb to survey) in order to assess its general shape, to "recognise its limits, to make a sketch and plant pegs at the corners of the angles" (F.I.C., 1882, p. 17). This preliminary reconnaissance allows the choice of the suitable method. It is necessary for the pupils to take "note on a well-made sketch of all the operations carried out", a framework which "must resemble as closely as possible the plan of the land being surveyed" (F.I.C., 1882, p. viii). David Puille also notes the importance of this drawing representing roughly the shape of the land because "by means of auxiliary lines (diagonals and perpendiculars), the surveyor can quickly decompose the polygon in the way that seems most favourable to his work: this decomposition also serves as a basis for him to act with order" (Puille, 1887, p. 164). The author indicates the following convention in the drawing of a sketch:

> it is customary to draw solid lines to represent the actual sides or contours of the lands; as for the auxiliary lines [...] used to divide the land [...] it is customary to use dotted lines. (Puille, 1887, p. 170)

Two further steps are then necessary to determine the area of a surface. The first takes place on the land itself, it consists of drawing straight lines, measuring segments and leading perpendiculars and parallels to given lines from points. The second operation, called "office work" (Hue & Vagnier, 1893, p. 311) consists of combining the lengths obtained in order to calculate the area of the land according to known principles, it is carried out in class. Here, we quickly pass over the measurement of the area of a land using only the surveyor's chain because it requires mathematical knowledge taught from the age of 13 (square roots). Indeed, it is a matter of dividing the surface to be measured into triangles and then measuring – with the chain – their sides (F.I.C., 1882, p. 19). The surveyor then uses the so-called Hero's formula to calculate the area of each triangle: denoting a, b and c the lengths of the three sides and p the semi-perimeter ($p = \frac{a+b+c}{2}$) the area of the triangle is equal to $\sqrt{p(p-a)(p-b)(p-c)}$. The total area is then obtained by summing the results obtained for each of the triangles.

The division of polygons on the ground is carried out in a way to limit the number of measurements and drawings. This allows to reduce the duration of the operations. The surveyor must also take into account the accessibility of the terrain. First, the area of a triangle is obtained by measuring on the ground one side and the height

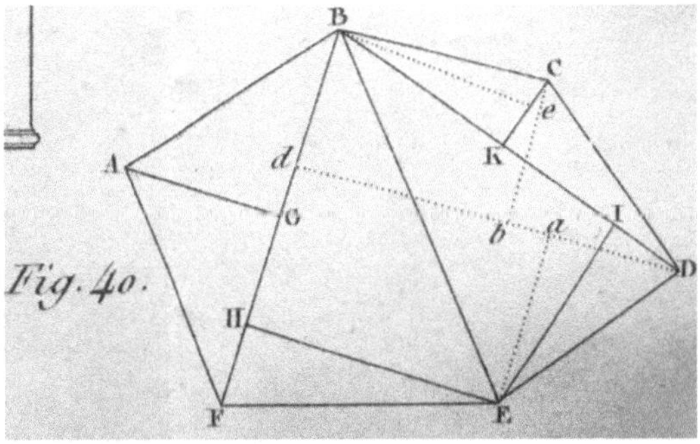

Fig. 7.7 Surveying methods (Lamotte, 1861, p. 21)

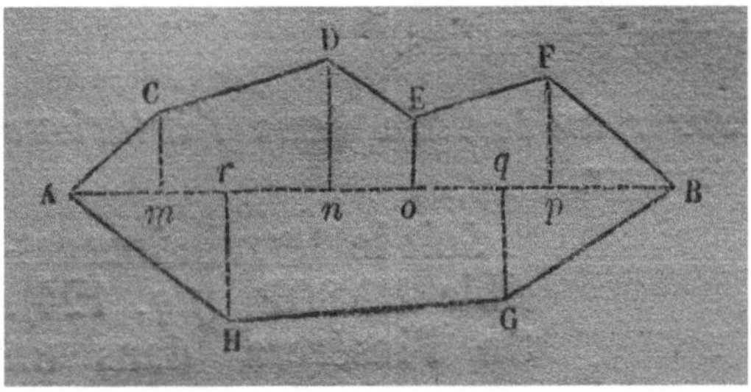

Fig. 7.8 Surveying methods (Eysséric & Pascal, 1862, p. 115)

from the opposite vertex (provided that a surveyor's square is available to draw a perpendicular) then by applying the formula for the area of a triangle: $\frac{\text{base} \times \text{height}}{2}$.

The other polygonal terrains are surveyed by three processes (F.I.C., 1882, p. 21):

1. decomposition into triangles (Fig. 7.7),
2. decomposition into right triangles and trapezoids,
3. or the use of inscribed or circumscribed polygons.

In the first method, the drawing of diagonals allows the plot to be divided into triangles, each of whose height is drawn at right angles. The height and base are then measured. The areas of each triangle are calculated and then summed to obtain the area of the land. In practice, this method is not the most used because it "requires the chaining of a large number of lines and the use of several different alignments to determine the heights of the triangles" (F.J.-O.P., 1875, p. 17).

The second method (Fig. 7.8), known as "by alignment or by director" (F.I.C., 1882, p. 21), is described as "the one most often used, the most convenient and the most expeditious" (Puille, 1887, p. 170). It begins by marking out a so-called director diagonal. Some precautions in its choice are however very useful. The director must be "easy to traverse" (F.I.C., 1882, p. 22). The surveyor must be able to see all the vertices of the land from this diagonal. It must be directed in the direction of the largest dimension of the land so that the perpendiculars to the diagonal are as short as possible. Thus, the marking out of the perpendiculars, also called ordinates, Frère Irlide Cazeneuve divides the surface into triangles and right-angled trapezoids

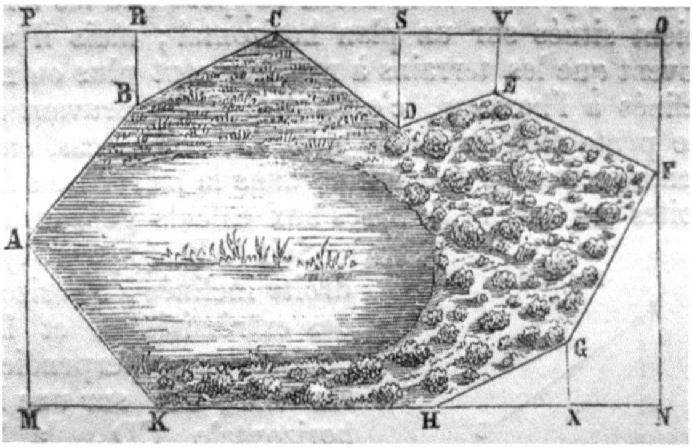

Fig. 7.9 Surveying methods (Eysséric, 1855, p. 115)

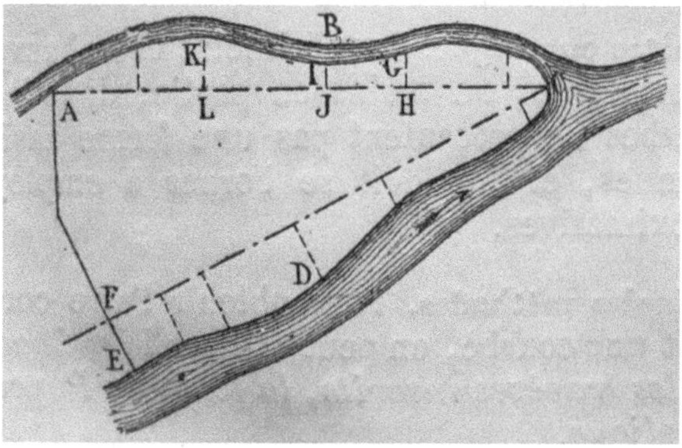

Fig. 7.10 Surveying methods (F.I.C., 1881, p. 36)

(F.I.C., 1882, p. 22). Their dimensions are then measured and transferred to the sketch. Often, the final calculations are arranged in a table of operations to facilitate data management (F.J.-O.P., 1875, p. 18).

The land, forest, marsh or field, to be measured can also be inaccessible. The square is then used to frame the surface with a large rectangle (Fig. 7.9) or any other easy-to-measure polygon. Each side of the polygon circumscribed to the inaccessible surface serves then as a director. The portions between the perimeter of the surface to be measured and the sides of the enveloping rectangle are then measured and the sum of their areas is subtracted from the area of the rectangle (F.I.C., 1882, p. 26).

The lands to be surveyed can also be bounded by curved lines. In this case, only an estimate of their area can be obtained by a compensation process or by the use of inscribed trapezoids (Fig. 7.10). The compensation process involves replacing the curved lines with broken lines chosen so that any portion subtracted is compensated by a portion added. The surveyor uses one of the previous methods to calculate the area of the polygon. If he is "skilful and conscientious" (F.I.C., 1882, p. 27), he obtains an approximate area of the land. The method of inscribed trapezoids consists of drawing segments close enough to the contour, named AC and FC in Fig. 7.10.

In this example, this thus defines two parcels located between the curve and these segments. Points are then placed on the curvilinear contours FC and AC. From these points, perpendiculars to AC and FC are raised. The more pronounced the curve, the more points must be numerous (F.I.C., 1882, p. 28). The curved part thus becomes very close to a broken line and the two parcels are divided into triangles and right-angled trapezoids. All that remains is to measure them and add them up to obtain a very good approximation of the desired area.

Areas: Freedoms of Procedures, Manipulation and Practical Problems

The measurement of the schoolyard's area is carried out over three years for pupils aged 10 to 12. Alongside this research, areas are taught following the recommendations of researchers from the secondary school group of a Mathematics Teaching Research Institute based in Poitiers. The progression revolves around the following three questions: How to compare areas? How to measure an area? How to calculate an area? (Groupe collège, IREM&S de Poitiers, 2010, p. 7).

Comparing surfaces according to their area, without resorting to measurement, is an activity aimed at "constructing the notion of area as an autonomous magnitude" (Perrin-Glorian, 1990, p. 33). At this stage, the area is not yet associated with a number that measures it. Surfaces are compared directly by inclusion and superposition or indirectly by cutting and reassembling. To compare surfaces, the pupil can, for example, cut out the proposed figures with his scissors. During this manipulation activity, he is free to propose a different cut than his classmates if it allows him to answer the question. These activities aim at several learning objectives "differentiate between area and surface, (surfaces of different shapes can have equal areas)" and "distinguish the area from the number" (Perrin-Glorian, 1990, p. 33).

The unit of measurement of areas is then introduced as an intermediary between two surfaces to be compared. The patterns of the tiling are of various shapes and sizes. Classic situations of counting tiles are proposed throughout the cycle. The tiling procedure then allows to measure the area of surfaces directly by counting the area units. They also give rise to the use of cutting and reassembling, "a key point in the elaboration of the concept of area" (Perrin-Glorian, 1990, p. 17). At this stage of teaching areas, the pupil's environment offers a wide range of possibilities for practical teaching of measurement. Marie-Jeanne Perrin-Glorian, in her study of French school curricula from elementary school and secondary school from 1887 to the reform of modern mathematics in 1970, noted injunctions to "operate in the yard, in the classroom, sometimes on the desk" and to "actually perform these operations with a graduated tape, accessible in practice to young children" (Perrin-Glorian, 1990, p. 15). From the first year, the pupil can thus evaluate the area of his desk using a square or rectangular template. The result is associated with a gesture of transferring the piece. This manipulation carried out in pairs allows to obtain a framing of the desk area. The use of square pieces one centimetre or one decimetre long also leads to questioning the choice of the usual area unit to use according to the surface to be measured. A similar approach is proposed by Sarah Selmer and her colleagues to address the difficulties of acquiring the concept of area: the use of the school garden space (Selmer et al., 2016, p. 4). In real gardening contexts, teachers suggest their pupils use physical square units of different sizes to measure areas.

Other familiar surfaces are also accessible through satellite images. The top view of a pond on a nearby leisure base, frequented by our classes, leads to using the hectometre square or the kilometre square to evaluate the area of this curved domain. The construction of a grid of the plane by a network is a long but necessary operation at the beginning of learning. Subsequently, a piece of transparent millimetre paper (Groupe collège, IREM&S de Poitiers, 2010, p. 10) is the tool used to determine the framing of the sought area. These individual searches on surfaces in the environment of our classes continue the construction of landmarks on areas and encourage not to neglect the estimation in favour of the search for the exact result. The units of measurement are not specified in advance so that each team can choose the most suitable unit. This point is particularly important to work on. Indeed, the examination of the results of the national assessments from CM2 to fifth grade showed significant rates of absence of unit in the pupils' responses (Moreira Baltar, 1997, p. 48).

The exploration of surfaces also contributes to the learning of calculating the area of composite shapes, an important application in daily and professional contexts (Lehmann, 2003, p. 531). These situations mobilise procedures of cutting and pasting, decompositions of a figure which Moreira Baltar also specified were not available "in complex situations, among a significant number of CM2 pupils and (to a lesser extent) among pupils" aged 12 in France (Moreira Baltar, 1997, p. 57). Similarly, large-scale pupil assessment in the United States and England showed, as early as 1980, that even the calculation of the area of basic composite surfaces posed difficulties (Lehmann, 2003, p. 531). In 2016, a national test in Australia in 9th grade, consisting of calculating the area of a square juxtaposed with a triangle, was passed by a third of the pupils.

The author relies on the large-scale assessment of pupils' ability to calculate the area of an L-shaped figure reproduced in Fig. 7.11. The researchers attributed the errors on Fig. 7.11 (a) to (d) to a lack of understanding of the concept of area, prioritising the application of formulas. The erroneous strategies of Fig. 7.11 (e) to (f) are linked to the pupils' lack of experience in reasoning about the geometric configurations of composites.

The situations proposed upstream of the surveying of the courtyard aim therefore, on the one hand, to familiarise pupils with a variety of configurations of composite shapes and, through their collaborative organisation, to encourage comparison between peers of the procedures for decomposing and recomposing a figure. These activities also require regular handling of measuring instruments and teamwork, necessary before surveying the courtyard.

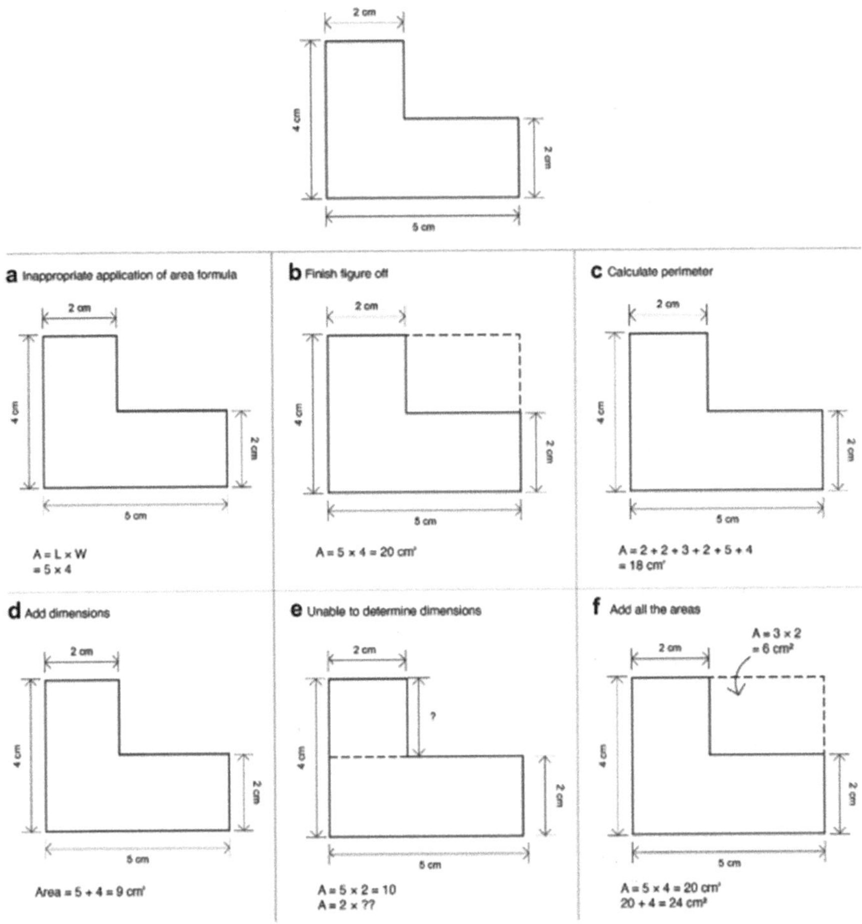

Fig. 7.11 Erroneous strategies (Lehmann, 2003, p. 533)

Making a Sketch of the Courtyard Then Cutting It Out

The surveying of the courtyard begins with the creation of a top-down view of the courtyard. This experience therefore leads the pupil to work in spaces of different sizes, between the classroom and the courtyard. The apprentice surveyor moves in a concrete space that is large in comparison to him. Depending on the configuration and obstacles present, his view of the courtyard is most often partial and requires him to later reconstruct the sketch from the information collected on the ground. The representation produced does not require the application of a scale but can be a relevant step before making a scale plan. Pupils aged 10 to 12 working in pairs or trios must also collaborate and describe their movements. Their exchanges mobilise a spatial lexicon that takes on its full meaning during this walk. The sketch drawn quickly freehand must be faithful to reality but only the main elements of the courtyard are retained. The pupils name the points to be able to communicate more easily about their figure. This sketch is then compared to a satellite image (obtained through cartographic visualisation sites Geoportal or Google Maps) in order to verify it before moving on to the next step. This consists of finding the most efficient method for determining the area of this complex surface, the culmination of learning about the concept of area. The initiative is then left to the pupil but he must face two constraints. The courtyard should only be cut into rectangles, squares or right-angled triangles, shapes whose area he knows how to calculate. In addition, he must make efficient cuts to avoid overly complicated manipulations on the ground.

Implementation in the Classroom: Stages of Surveying

In this section, we describe the changes made to our usual teaching of the concept of area in order to link it with the different stages of measuring the yard. We then outline the process of surveying and the results obtained following the introduction of this guiding question into our teaching.

Preliminary Reflections on Classroom Experiments

A Practical Teaching of the Concept of Area

Measuring a schoolyard is a problem-solving situation that comes after many lessons. It can therefore be announced as soon as areas are discussed in class, as the objective of the lessons on this subject. In order for the pupils to be focused on the investigative approach, it seemed necessary to us that they had become accustomed in advance to taking measurements with decametres or calculating the areas of concrete spaces.

Thus, once the formula for the area of the rectangle has been established using the grid for pupils aged 10 and over, the classroom again offers an excellent exercise for calculating the areas of the present rectangles. The pupils, grouped in teams of four, have a decametre and must take measurements on the rectangular surfaces assigned to them. The distribution ensures that several teams measure the same surface. Answering this question involves knowledge from several disciplines and areas: handling rules or decametres, effective cooperation within a team, conversions of length units or area units, the use of a formula and operations on decimal numbers. At the end of this work, the results of which can be added to the lesson, the pupil has a wide range of reference areas expressed in units ranging from square centimetres to square metres.

The formula for the area of the right triangle is then specified and, by cutting and reassembling without scissors or glue, it also becomes possible to determine the areas of more complex polygonal shapes. Alongside traditional exercises, we can again propose estimating the areas of neighbouring zones of the establishment from satellite images. The procedure is left to choice. We noticed that pupils were more involved in these exercises, which they perceived as more concrete than traditional applications. They are curious to know the result and draw sufficient motivation to tackle this problem despite the complexity of the concepts involved. We can then guide them towards a division of the surface that is most efficient for calculations. It is particularly more practical to limit the number of arbitrary triangles.

The problem-solving situation of surveying the yard can then be addressed in four stages, spread over several weeks, over three years, alternating between classroom and field. The problem-solving can be introduced by a historical slideshow, based on the elements developed in the first two parts, thus linking the history and mathematics programmes.

First Step of Field Surveying: The Schoolyard Sketch

In the fine weather, the class begins its investigations in the field. During this first stage, the pupils go to the schoolyard to construct a sketch of it. The class is divided into teams of two or three pupils and the instruction is to draw a sketch of the schoolyard. The equipment consists of a cardboard support or a rigid book, a pencil and an eraser. The goals of this field stage are to develop spatial vision. The pupils' productions can be very diverse and their sharing will vary according to the classes.

Second Step of Field Surveying: Cutting the Yard Sketch

The second phase is carried out in the classroom. The canvas is validated by the pupils and then cut into simple shapes whose area we know how to calculate using a formula. Projecting a satellite image or an extract from the land registry can facilitate the debate if one of the sketches obtained does not win the pupils' approval. This step is crucial. If the pupils agree that they cannot calculate the area of this figure directly, it is more delicate to choose a cut-out that allows for simple manipulations. Relying on a freehand figure also means that the other drawings will be done freehand. Right angles will not be drawn with a set square but indicated by codings. This canvas must be checked on the ground before taking measurements in the yard.

Third Step of the Surveying on the Ground: The Measurements in the Yard

The class returns to the field again to carry out the constructions with the surveyor's square and the necessary measurements with the decametre. During this manipulation, many numerical data must also be handled. Two hours are needed to familiarise oneself with the surveyor's square, divide the schoolyard into rectangles and right-angled triangles and then measure the lengths of the segments useful for calculating areas. The surveyor's squares are placed at the vertices of the rectangles and right-angled triangles which divide the field. The apprentice surveyors must imagine the visual rays joining these points. Each point is marked with chalk on the ground and the pupils measure the distances between these points and then record them on a chart accompanying the plan (Fig. 7.12). The exercise also mobilises the pupil's ability to collect and classify data. Obtaining a large number of measurements for each segment then allows to discard possibly false measurements. The teacher, accompanied by one or more pupils, will carry out during or after the session some checks of these measurements in order to avoid possible errors.

Fourth Step of the Surveying on the Ground: The Calculation of the Area of the Yard

Final stage of the research, the apprentice surveyors calculate the total area of the yard from the data collected in the field. The pupils must mobilise the correct formulas to determine the areadecs of the figures obtained. This session also requires skills related to calculations on decimal numbers. To maintain the pupils' motivation if they have already measured the surface the previous year, this session can be differentiated. The calculations can be done partly by hand and with a calculator for first and second year pupils and with a spreadsheet the following year.

Report of Experiences Between Classroom and Field

Old Manuals as Sources of Curiosity and Motivating Manipulations of Instruments

The problem situation was introduced by a historical presentation for the 10 and 11 year old pupils. Following this twenty-minute presentation, the pupils were able to handle old or recent measuring instruments as well as primary school textbooks from the Third Republic. While flipping through the pages of these old books, many recognised the illustrations present in the slideshow and especially the instruments. They questioned their teachers about the materials needed for their manufacture. They held them in their hands and made sightings with the surveyor's square.

The second-year pupils, having already measured the school yard the previous year, were more reactive than the younger ones discovering the problem. They pointed out that they had not seen the square of a brass surveyor, the previous year.

The teaching of areas in connection with the environment has shown a great motivation of the pupils during sessions requiring manipulations. In the third year, more than twenty rectangles were measured in the class and the teacher even had to limit the number of figures so as not to devote too much time to this activity. The problems of measuring complex surfaces from satellite images also called for a wide variety of procedures. Some used their gridded tracing paper or constructed a grid of the plan and then counted the squares. Others cut the surface into rectangles and triangles, obtaining very varied cut-outs.

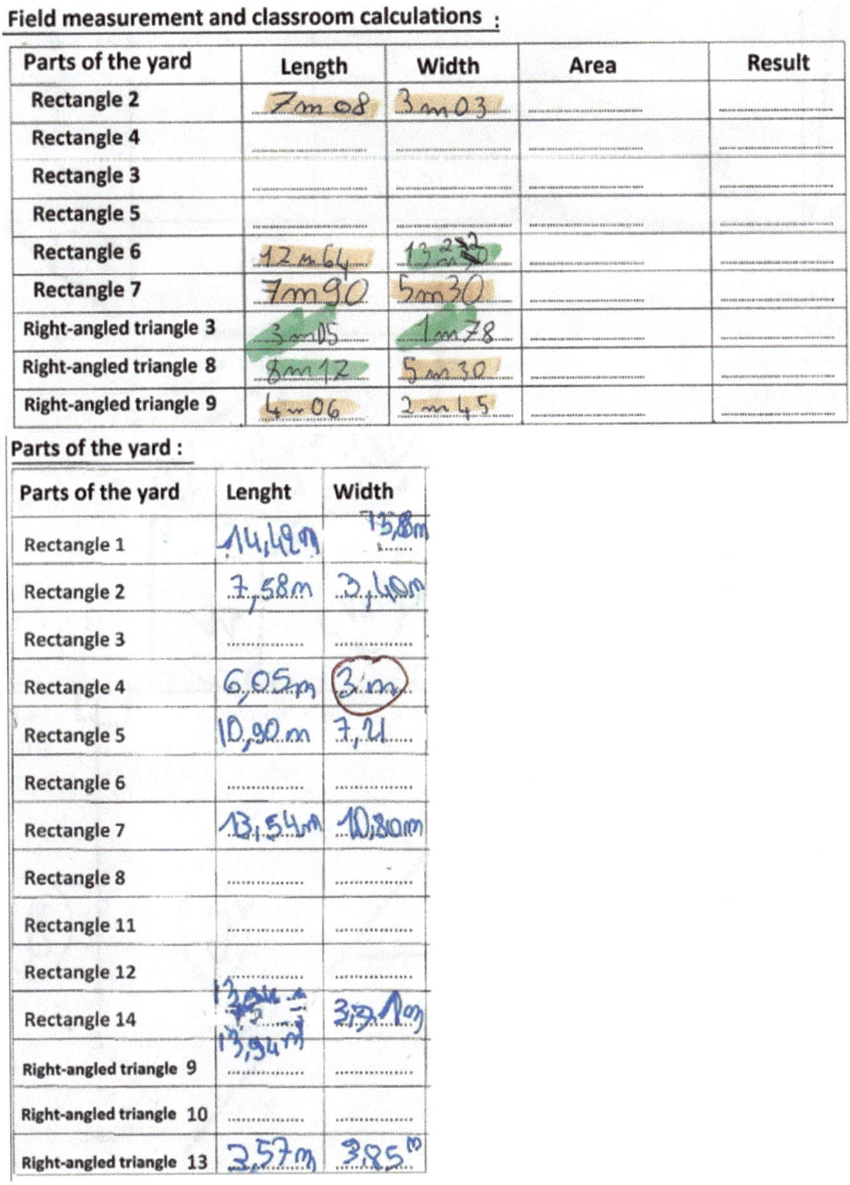

Field measurement and classroom calculations :

Parts of the yard	Length	Width	Area	Result
Rectangle 2	7m 08	3m 03		
Rectangle 4				
Rectangle 3				
Rectangle 5				
Rectangle 6	12m64	13m 23		
Rectangle 7	7m90	5m30		
Right-angled triangle 3	3m05	1m78		
Right-angled triangle 8	8m12	5m30		
Right-angled triangle 9	4m06	2m45		

Parts of the yard :

Parts of the yard	Lenght	Width
Rectangle 1	14,49m	13,8m
Rectangle 2	7,58m	3,40m
Rectangle 3		
Rectangle 4	6,05m	3 m
Rectangle 5	10,90 m	7,21
Rectangle 6		
Rectangle 7	13,54m	10,80m
Rectangle 8		
Rectangle 11		
Rectangle 12		
Rectangle 14	13,94m	3,31m
Right-angled triangle 9		
Right-angled triangle 10		
Right-angled triangle 13	3,57m	3,85m

Fig. 7.12 Measurements obtained at school and college

The Exercise of the Sketch

When making the sketch, the 10 to 11 year old pupils most often sat on benches to draw. Gwenaëlle and her friend even lay down on the ground. For the sake of realism, Noémie used her two hands to report an obtuse angle on the sketch. In their secondary school, the 12-year-old pupils naturally positioned themselves in the forum overlooking the courtyard. We were able to record the activity of all without exception. Théo, usually not very invested and struggling in the subject, took charge of the pair's work.

As some groups had not finished at the end of the half-hour, we gave them the opportunity to complete their work and to hand it in the next day. Cooperation within the teams was very variable but we observed that the pupils who were doing this activity for the second or third year were very comfortable and were able to quickly draw a sketch. The analysis of the canvases also revealed that no pupil had tried to orient his plan in relation to the cardinal points. They naturally positioned themselves facing the courtyard from the door we had exited.

The second phase, carried out in the classroom, started with a half-hour of classroom debate, led by the teacher. In the third year, in the secondary school, all the sketches were projected and compared to a satellite image of the courtyard. The orientation was therefore carried out according to the cardinal points. The sketch representing a front view rather than a top view, made by a pair, was quickly discarded. This discussion thus provided an opportunity to reinvest the vocabulary of views. Criteria for respecting proportions or care taken in the drawings were proposed to discard or retain the solutions. Finally, the sketch made by Anna's team was chosen. To make this sketch, this pair took measurements step by step in the courtyard. At the school, the scribes had previously gone over their production in black felt pen before displaying it on the board. The pupils first tried to define criteria to designate the best sketch: correspondence with reality (subjective point), respected dimensions, quality of the pencil stroke, details, codings. Merwan also indicated that the sketches were not all oriented in the same direction. The pupils chose the portrait mode, with the gate placed at the top. Some sketches were then eliminated because they did not meet the initial instruction. One included the playground and the roofs of the school even though they were not requested. Another was rejected because the gate appeared much too large. Oscar specified: "this will lead us astray". The discussion then turned to the choice of details to keep. Some could hinder the calculation of the area of the courtyard but others like the tree and the benches allowed to find one's way during the measurements. Finally, only two plans remained and the one with the most faithful representation of the gate was kept. The two discussions held at the school and at the secondary school led to the observation that comparing the surveys of the courtyard to a satellite image would be of great help. The criteria retained by the class are the same as at the school: type of view, correspondence with reality, respected dimensions, quality of the pencil stroke, details, codings. Not surprisingly in retrospect, the retained sketch was made by a team whose one pupil had already constructed a plan of the courtyard of her school in another region the previous year!

Following the validation of the sketch, a geometric task of dividing the representation of the courtyard was undertaken in order to calculate its area. The second and third year pupils easily engaged in the activity. The first year pupils encountered more difficulties, which cannot be explained solely by the fact that the activity was at the beginning of learning the concept of area. Indeed, it seems that the recognition of the geometric figures involved and the concept of a right angle had not been well assimilated previously. Often, the pupils did not start from the outline of the courtyard but drew rectangles and triangles inside it. Sarah, for example, was worried about a too high number of squares inside her figure. The resolution of this question of measuring the area of a complex surface, requiring cutting and reassembling methods, probably came a bit early but it was difficult to exclude the first years from a project that motivated them. All the pupils therefore made cuts of the sketch. We then took the initiative and selected one of the productions.

Measurements in the Courtyard

The third step led the pupils onto the field. Each class had five surveyor's squares, about ten pegs and about ten decameters. The Appendix details the construction methods of a square (if necessary, the physical education and sports or technology colleagues have decameters). The session was introduced by a ten-minute presentation in class. The pupils suggested numbering the figures and using a table to classify all the data collected in the field.

This allowed the construction of a document containing the selected sketch and a table classifying the dimensions to be measured: length and width of rectangles, sides of the right angle for right triangles. Once in the courtyard, the apprentice surveyors were instructed to orient their plan to understand where the areas to be measured were located.

Several volunteer teams were therefore sent to rectangular areas, bounded by buildings or the playground. About twenty pupils stayed to make the first placements of squares with the teacher. The latter assigned roles to his pupils: distribute the chalks to make the markings on the ground, hold the decameters or squares, etc. We noticed during the session that it was important to specify again how to take a measurement with a decameter.

This must be placed on the ground, horizontally. It is also necessary to remind pupils that they should not perform any calculations at this stage because, otherwise, the class would not take enough measurements in the allotted time. Aligning stakes is a fairly long phase because their position must be perfectly vertical (Fig. 7.13). The measurement of two areas is however sufficient to assimilate the procedures and the apprentice surveyors then dispersed in the courtyard.

Geometric knowledge about polygons allowed pupils to speed up their work. In particular, the properties of the sides of rectangles, adjacent figures 3 and 6 or 11 and 12 of the plan of the school courtyard (Fig. 7.14) also avoided certain measurements. With one exception, the pupils, very autonomous during the session, helped each other between teams. The difficulty in "seeing" all the pupils is a new and somewhat destabilising situation for the teacher. However, our physical education and sports

Fig. 7.13 Aligning the markers at a right angle; measuring lines with a decametre (Photo: Marc Troudet)

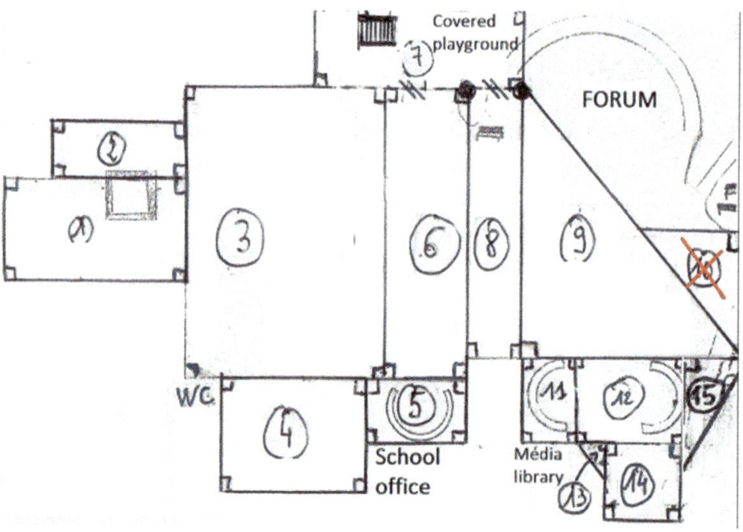

Fig. 7.14 Sketch of the schoolyard

colleagues (and therefore our pupils) are used to this class management. We plan to turn to them for our next field experiences.

New Tools—Spreadsheet and Geoportal—For Calculating the Area of the Yard

After two hours of constructions and measurements carried out in the yard, we collected the data and kept the most relevant results by performing a verification using the distance and area measurement tools of the Geoportal site. The verification of four measurements, difficult to control on the satellite view, was carried out the next morning. We noticed that triangles 13 or 15 were little measured and will require our vigilance for the following years. The investigations had indeed started in the zones 3 and 4, near the pupils' gathering place. It would be appropriate to impose areas to measure on each team to avoid this situation.

A few days later, the schoolchildren carried out measurements in their yard (Fig. 7.15). They made the same observations.

Among the apprentice surveyors, twelve second-year pupils and five third-year pupils, from this school group, had carried out the survey of their yard at the end of the previous year. These pupils were the most autonomous and enthusiastic throughout this measurement session.

The fourth and final phase of our work was requested on many occasions by the 12-year-old pupils because they wanted to know the result. A forty-minute session was devoted to calculations, in the form of individual work for this level.

We distributed to the first-year pupils a plan of the yard constructed on a grid paper (Fig. 7.16), the lengths having been rounded to the nearest meter so that they could calculate an estimate of the area of the yard. As was predictable, a few schoolchildren had trouble with the use of formulas and understanding the table. After their teacher's intervention, they managed to perform their calculations.

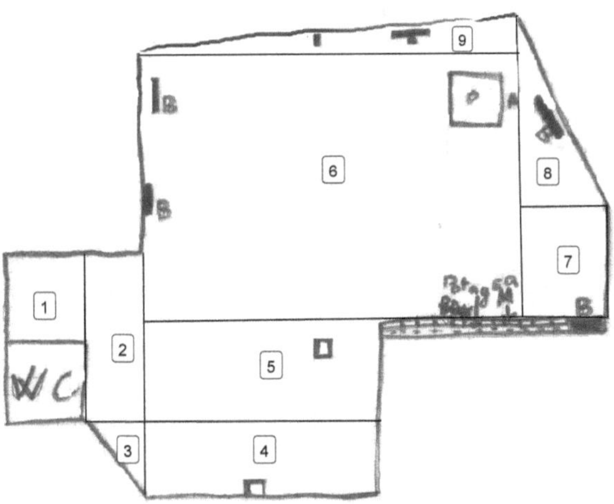

Fig. 7.15 Sketch of the school playground

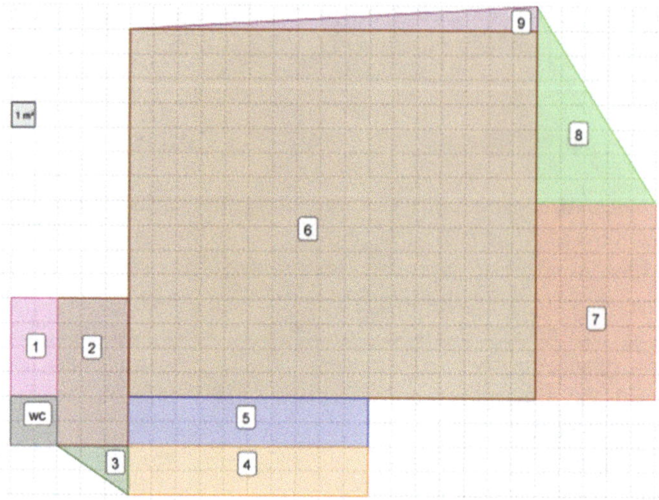

Fig. 7.16 Sketch of the gridded playground for the 10-years old pupils

Counting the full squares was done by enumerating them one by one or by using multiplications, an approach that we wanted because it motivates the introduction of formulas in second year. For this class level, the main difficulty was to use the formula of the area of the right triangle. The division by 2 was forgotten several times. The second-year pupils applied the formulas of the areas of the rectangle and the right triangle to determine the areas of all the zones. They then added the results obtained to calculate the total area of the yard.

Following the surveying session, several third years were surprised not to have surveyed the area containing the forum. The height of the stands indeed prevents the use of instruments. They asked us to calculate the total area of the yard, thus including the surface impossible to survey (Fig. 7.14). The desire to determine the total area of the yard led us to divide the yard into two zones. The calculation of the area measured with the decameter and the square was carried out using the spreadsheet. As at school, a few pupils forgot the division by 2 when calculating the area of the triangle. The surface of the inaccessible area was obtained using the measure a surface tool available on the Geoportal site (Appendix 2). The pupils did a search for their school and then measured the area of the forum zone. They then reported this value in the spreadsheet file. At the end of this one-hour session in the computer room, the apprentice surveyors obtained an estimate of the total area of the yard.

This result was inserted in their course notebook, next to the reference areas calculated throughout the year.

Conclusions

The observations that we were able to make during the two years of classroom experiences lead us to draw conclusions of various kinds. Proposing the same problem situation to a class over three years has not been a hindrance to pupil motivation. The differentiation previously exposed even allows this work in a double level and can therefore adapt to any local progression. It allows everyone to achieve a satisfactory result in solving a problem that questions and concerns them. In the first year, it can even serve as an introduction to the use of new methods for calculating the areas of usual figures: formulas. The change of school yard with the transition to secondary school and the use of calculation software or cartographic visualisation sites also re-motivate the third years. During this work, the most fragile pupils have progressed in their mastery of procedures and the most comfortable pupils have practised on more ambitious situations, comparable to challenges. Anna found it "funny" to discover new techniques for more complex figures. Amine and Jibril noted that a larger yard increased the difficulty and differed from primary school. Raphaëlle, Théo and Jibril appreciated the collective resolution. For Maxime, "it was good, because we learned to cut out figures and understood what it was for". Everyone noted the interest of using a calculation software and many pupils comfortable in the discipline appreciated solving a problem initially considered difficult or even out of reach, with the knowledge they had.

The spreading out of activities over time seems to have pleased the pupils. Anna preferred this version to a resolution carried out over half a day in the second year. As soon as the investigations started, several third years kept asking for the next part. The fact of asking a question arising from the pupils' concrete environment seems to us to be decisive in their motivation. However, back and forth with the field are decisive for the problem to arouse full engagement. Thus, all third-year pupils managed to break down the canvas into rectangles and right-angled triangles measurable on the ground. We then proposed to a second third -year class, following the same teaching on areas but without going into the yard, to cut out the sketch of the yard made by their classmates. More than a third of the pupils did not perform this cutting correctly. Figure 7.17 shows an example of incorrect production. The surface is cut into triangles and these overlap, so this cutting does not allow the calculation of the area. Other pupils cut the sketch into triangles without overlap but, this time, the triangles were not right-angled and seemed difficult to measure on the ground. We hypothesise that the absence of the intermediate session in the yard made this exercise too "artificial".

The surveying of the yard is only made possible by the manipulation of instruments. This motivates pupils who are dropping out to start the activity and also leads to many exchanges between peers about the methods to use to solve the problem. Moreover, we believe that it plays a role in the distinction between area/perimeter. Indeed, all operations are carried out inside the surface to be

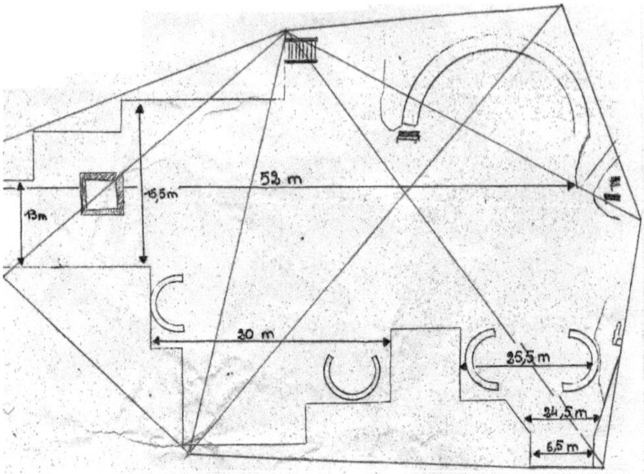

Fig. 7.17 Example of cutting the canvas of the courtyard

measured and not on its contour. It is also possible to refer to it during subsequent area calculations by evoking the memory of the experience in the field. Lisa's example has reinforced this opinion. This third year pupil's assessments had shown that she confused area and perimeter. When measuring the sides of right-angled triangle 13 (Fig. 7.14), she wanted to measure the hypotenuse. Her two classmates dissuaded her by re-explaining the method. We questioned Lisa several weeks after the measurements made in the yard. She had to calculate the area of a right-angled triangle with sides three, four and five drawn freehand. Lisa first started by adding the sides of the triangle as if to calculate the perimeter. The simple reminder of the triangle measured on the ground led her to realize that she was not calculating the right quantity.

She finally performed the correct calculations. The method is therefore not yet assimilated but the mental image associated with the exercise carried out in the field allowed Lisa to correct her mistake.

If the use of decametres helps to consolidate knowledge about units of length, the manipulation of a surveyor's square links the act of constructing a perpendicular on the ground to the calculation of the area. In our classes, measurements, called toises under the *Ancien Régime* in France, are quite common and often illustrate the calculation of a figure's perimeter. During the reform of units of measure, surveying did not find a replacement while the arpent, a unit of area used under the *Ancien Régime* in France, was replaced by the square metre. We could substitute it with the term "squaring" with an expanded meaning so that the word better reflects the action to be performed and supports the learning of the concept of area. The measurement of the area of a schoolyard is not reserved for mathematics alone. Our experiences have also highlighted interdisciplinary work paths with French, history and geography and our research will continue in this direction. Surveying appears to us as one of the pillars of the bridges that we build both between disciplines and between primary and secondary education.

Acknowledgments This chapter was written by Marc Troudet, mathematics teacher, with the help of Saida Boué and Pauline Mirmand, primary school teachers (IREMI de Grenoble and IREM de Lyon), France.

Appendix 1: Concepts and Skills Addressed in This Chapter

Concepts Covered

- Orienting oneself and moving through space:
  - Identify, describe, or carry out movements in order to create and read a sketch of the schoolyard.
  - Use vocabulary to describe positions and movements on the sketch and within the schoolyard.
- Recognizing and using some geometric relationships : Make drawings that illustrate perpendicular or parallel relationships between lines and segments.
- Comparing, estimating, and measuring geometric quantities (here, areas):
  - Distinguish between area and perimeter of a surface.
  - Determine the area of the schoolyard using simple tiling (square, rectangle, triangle) or by applying a formula.
  - Estimate the measurement of an area using various methods.
- Solving problems involving geometric quantities using whole numbers and decimals:
  - Solve a land surveying problem.
  - Use different units of measurement simultaneously and perform conversions (multiples and submultiples of the square meter).

Skills Developed

- **Search and research skills**: Engage in a mathematical approach by mobilizing previously encountered tools or procedures, and by constructing reasoning appropriate to a new situation, such as determining the area of a schoolyard. Test and explore multiple problem-solving strategies.
- **Computational skills**: Perform calculations with decimal numbers, either exact or approximate, using appropriate strategies or techniques (mental math, horizontal layout, or written operations). Verify the plausibility of results. Use a calculator to obtain or check a result when necessary.
- **Mathematical communication skills**: Progressively use appropriate mathematical vocabulary and/or notations to describe a situation and to present an argument when comparing different sketches of the schoolyard. Explain one's reasoning or approach, understand others' explanations, and engage in argumentation during mathematical discussions.
- **Mathematical representation ability**: Use appropriate tools to produce a sketch of the schoolyard.
- **Modeling**: Identify real-life situations that can be modeled through geometric relationships (alignment, parallelism, perpendicularity).

Appendix 2: Material Considerations

Method of Constructing a Simplified Surveyor's Square, Extracted from Ancient Works

The availability of instruments is not a new issue. It arises very early in schools and some textbook authors offer solutions to those who do not have instruments. For example, the *Notions de géométrie pratique et d'arpentage* by Joseph Eysséric provide an example of construction for those who do not have a square. The author points out that the material issues of teachers are not taken into account by other works. His work is therefore written in the following spirit:

> to do practical geometry with the instruments that everyone has, to replace expensive and complicated instruments with very simple instruments that one can build oneself. (Eysséric, 1855, p. 56)

He then proposes to

> build a wooden cube (Fig. 7.19 (left)) seven or eight centimetres on the edge. Two very fine saw lines, perpendicular to each other, are drawn on the top face and descend to half the thickness of the wood, thus forming the four pinnules of the square. The bottom face is

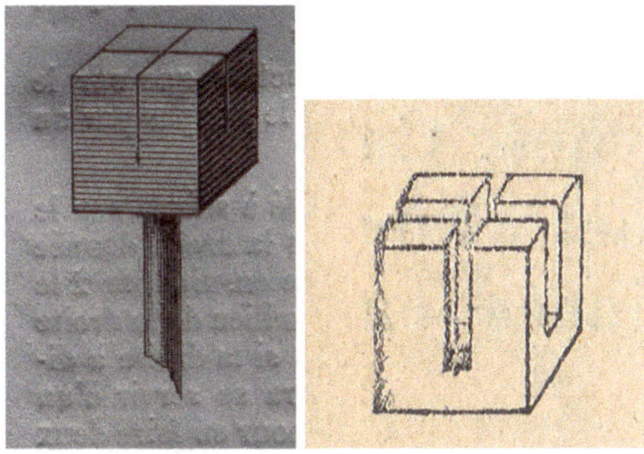

Fig. 7.18 Examples of surveyor's square (left) (Eysséric, 1855, p. 55) (right) (Hue & Vagnier, 1893, p. 314)

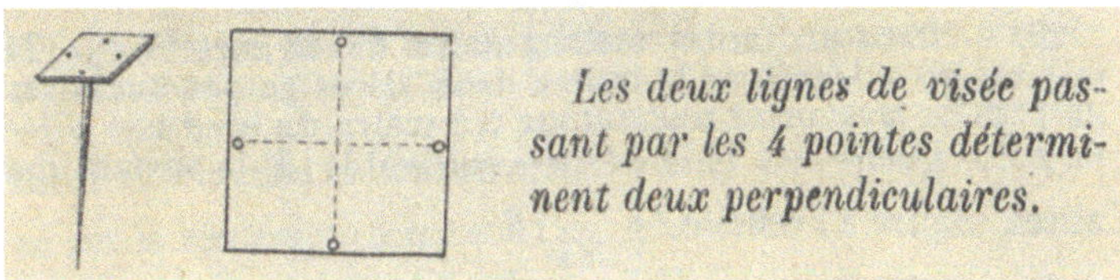

Fig. 7.19 Another example of a surveyor's square in (Pécherat, 1897, p. 28) (English translation) "The two lines of sight passing through the 4 points determine two perpendiculars"

> pierced with a cavity that allows the instrument to be placed on a stick one metre fifty in length. (Eysséric, 1855, p. 55)

Joseph Eysséric adds that by giving this wooden square ten centimetres instead of eight, one would have the cubic decimetre, "so useful in the study of metric teaching" (Eysséric, 1855, p. 55).

T. Hue and N. Vagnier propose, in their *Géométrie plane, arpentage et levé de plan*, to

> replace the surveyor's square with a simple wooden cube in which two perpendicular saw lines would have been made, passing through the axis of the cube parallel to the faces and up to about two thirds of its height (Fig. 7.18 (right)). A hole will allow the instrument to be fixed on a stick. (Hue & Vagnier, 1893, p. 314)

In *La géométrie pratique, arpentage et cubage à l'usage des écoles primaires, des cours d'adultes et des familles*, J. Pécherat—presumably a teacher according to the cover page—indicates at the beginning of the lesson dedicated to surveying that the square has a relatively high cost. He advises the following construction method (Fig. 7.19):

> Take a square board twelve to fifteen centimetres on each side, draw through the middle, using a set square, two perpendicular lines. On these lines and as close as possible to the edges, plant four small pins. Fix the instrument to a stake of about one metre fifty. (Pécherat, 1897, p. 28)

Fig. 7.20 Surveyor's square used by the pupils (made by the AMAFI association—Toy Ark, from Fontaine, a town in the French department of Isère, Photo: Marc Troudet)

Method of Construction of the Surveyor's Square Used in Class

In Fig. 7.20, you can see a photo of one of the surveyor's squares used by our pupils to carry out the surveys of a courtyard. In a right wooden block (10 cm × 10 cm × 5 cm), two slots of length two and a half centimetres are cut, perpendicular in the middle of the sides of the top square face. The block is then mounted on a light tripod. Three to five squares and about ten decametres allow a class to be equipped. By superimposing two surveyor's squares, a dm^3 is also obtained and this assembly can be used to illustrate the lesson on volume units.

Appendix 3: A Cartographic Visualisation Sites and Measurement Tools

1. Measure an area with Google Maps:
   Go to the site https://www.google.com/maps.

   - Use the satellite view to locate the area to measure
   - Right-click on an edge of the surface to measure
   - Follow the perimeter of the surface and go around
   - The total distance and the area of the land are displayed at the bottom of the screen

   In the computer room or on a digital tablet, students can then redraw the courtyard on Google maps or another map-viewing site such as Geoportail, then compare the area found with the site with that found in the field.

References

Boucheny, G., & Guérinet, A. (1912). *La géométrie au cours complémentaire*. Larousse.
Buisson, F. (1887). *Dictionnaire de pédagogie et d'instruction primaire*. Hachette.

Buisson, F. (1911). *Nouveau dictionnaire de pédagogie et d'instruction primaire*. Hachette.

D'Enfert, R. (2003a). Inventer une géométrie pour l'école primaire au XIXe siècle. *Trema, 22*, 41–49.

D'Enfert, R. (2003b). *L'enseignement mathématique à l'école primaire de la Révolution à nos jours. Textes officiels. Tome 1 : 1791-1914*. INRP.

Eysséric, A., & Pascal, J. (1862). *Traité de géométrie théorique et pratique* (11th ed.). Dezobry et Tandou, Librairies-éditeurs

Eysséric, J. (1855). *Notions de géométrie pratique et d'arpentage à l'usage des élèves qui passent de l'école primaire aux champs ou à l'atelier*. Langlois et Leclercq.

F.I.C. (1881). *Arpentage, levé de plans et nivellement*. Mame et fils & Poussielgue Frère, Tours & Paris. (F.I.C. = Frère Irlide Cazeneuve)

F.I.C. (1882). *Manuel d'arpentage pour les écoles primaires*. Mame et fils, Tours. (F.I.C. = Frère Irlide Cazeneuve)

F.J.-O.P. (1875). *Arpentage, levé des plans et nivellement*. Mame et fils & Poussielgue Frère, Tours & Paris. (F.J.-O.P. = Frère Jean-Olympe Paget)

Frères des écoles chrétiennes. (no date). *Géométrie Cours moyen*. Mame et fils & Poussielgue Frère, Tours & Paris

Frères des écoles chrétiennes. (1899). *Manuel d'arpentage pour les écoles primaires*. Mame et fils & Poussielgue Frère, Tours & Paris.

Groupe collège, IREM&S de Poitiers (2010). *Enseigner les mathématiques en sixième à partir des grandeurs: les aires*. IREM&S de Poitiers.

Hue, T., & Vagnier, N. (1893). *Géométrie plane, arpentage et levé de plan*. Delagrave.

Laisant, C.-A. (1906). *L'initiation mathématique*. Hachette.

Lamotte, L. (1861). *Traité élémentaire d'arpentage, nouvelle édition*. Hachette et Cie.

Lehmann, T. (2003). Learning to measure the area of composite shapes. *Educational Studies in Mathematics, 112*, 531–565.

Moreira Baltar, P. (1997). À propos de l'apprentissage du concept d'aire. *Petit x, 43*, 43–68. https://irem.univ-grenoble-alpes.fr/medias/fichier/43x6_1568724517582-pdf

Pécherat, J. (1897). *La géométrie pratique, arpentage et cubage à l'usage des écoles primaires, des cours d'adultes et des familles*. Badel

Perrin, D. (2005). *Mathématiques d'école, nombres, mesures et géométrie*. Cassini.

Perrin-Glorian, M.-J. (1990). L'aire et la mesure. *Petit x, 24*, 5–36. https://irem.univ-grenoble-alpes.fr/medias/fichier/24x1_1570438041435-pdf

Puille, D. (1887). *Cours complet d'arpentage élémentaire, théorique et pratique à l'usage des divers établissements d'instruction* (18th ed.). Fouraut et fils

Rapet, J.-J. (1855). De la direction à donner par les instituteurs à leur enseignement. *Bulletin de l'instruction primaire, 3*, 122–124.

Selmer, S., Valentine, K., Luna, M., Rummel, S., & Rye, J. (2016). How can we best use our school garden space? Exploring the concepts of area and perimeter in an authentic learning context. *Australian Primary Mathematics Classroom, 21*, 3–10.

Marc Troudet is a mathematics teacher in middle school and a member of the Institute for Research on Mathematics Education (IREM) in Grenoble. He has organized numerous practical geometry experiments, using old measuring instruments, in primary and middle schools.

Part III
Space and Geometry

Chapter 8
The Geometry of the Notebooks of Leonardo da Vinci

Marc Moyon

> *Iron rusts from disuse, stagnant water loses its purity and in cold weather becomes frozen; even so does inaction sap the vigors of the mind.*
>
> Leonardo da Vinci, *Codex Atlanticus,* fol. 285v

Abstract This chapter proposes teaching plane geometry to children aged 9–11 through the sketches of Leonardo da Vinci (1452–1519) found in the *Codex Atlanticus* (Biblioteca Ambrosiana, Milan). Drawing on Renaissance scientific questions, the approach focuses on constructing geometric figures, particularly by reproducing Vinci's works. The pedagogical method, based on mathematics didactics research, encourages a progressive learning path from the geometry of shapes to the geometry of points and their properties, with an emphasis on alignment. This historical immersion enriches pupils' understanding and fosters deeper geometric learning.

Keywords Leonardo da Vinci · Geometry · Geometrical construction · Rules and compass · Renaissance · Original documents in the classroom

In this chapter, we propose activities for reproducing complex figures, all composed of squares and circles. We thus wish to work on constructions using the ruler (non-graduated) and compass, at the heart of geometric learning (from 8 years old), particularly as a preferred tool for transferring length. In addition to the use of specific vocabulary, the construction of these figures involves, among other things, the concepts of points of intersection, parallelism and perpendicularity.

For this, we have chosen to start from the geometric works of Leonardo da Vinci. Most pupils know him as a painter or inventor, but they are often far from imagining

M. Moyon (✉)
Institut XLIM UMR CNRS 7252, IREM de Limoges, Université de Limoges, Limoges, France
e-mail: marc.moyon@unilim.fr

M. Moyon et al. (eds.), *Teaching Mathematics Through Historically-Based Activities*, History of Mathematics Education,
https://doi.org/10.1007/978-3-032-16990-7_8

all his mathematical works. We first sketch, in broad strokes, the life of this artist as famous as he is enigmatic. Thus, we describe some elements of the historical and scientific contexts of the early decades of the Renaissance during which the actors draw from the mathematics of Antiquity and the Middle Ages to develop their own activity. We then present the pedagogical devices as we have imagined and implemented them in classes, accompanied by the necessary didactic analyses. We thus account for, as precisely as possible, the different works of the pupils (between 8 and 11 years old). Finally, in a last part, we present possible extension paths in interdisciplinarity.

Historical Context

Leonardo da Vinci (1452–1519): Some Biographical Landmarks

To write this part, we used above all works written by Ackermann (1998), Denizeau (2016), and Isaacson (2017).

Leonardo was born in 1452 in Vinci, a small town in Tuscany. An illegitimate child, his father not recognising him, he will not have the right to enrol at university. He is a self-taught individual who educates himself from the books of his personal library, or in contact with contemporary scholars. A few years before 1470, he becomes an apprentice in the *bottega* [workshop] of the famous artist Andrea Verrocchio, in Florence. At 20, he is officially a member of the guild of Florence painters. It is also in Florence that he invents his first machines and concerns himself with the perpetual movement of water. A decade later, in 1482, he presents himself as an artist and engineer in Milan where the Duke of the city, Ludovico Sforza, will protect him within his court. In Leonardo's time, to practice his art and be sheltered from need, it is usual and comfortable to be the protege of a rich and powerful man. This is patronage. Mathematicians are exactly in the same situation as artists; it is—apart from the status of teacher, at university or within specific professional corporations—the only way to have a guaranteed income. Of course, it must meet the needs and other demands of his patron but it is also the price of freedom. In Milan, in addition to his work in painting, he studies music or the transformation of movement in order to develop machines and other automata to animate the Duke's parties. He conducts his first studies on the horse to create a statue. In 1490, he is recognised in Milan as an eclectic artist.

It is also at this time that he begins to take an interest in mathematics. Indeed, it is in Milan, in 1496, that he meets Fra Luca Pacioli (15th/16th century), an important mathematician to whom we owe notably the *Summa de aritmetica geometria proportioni et proportionalita*, a work that covers the main areas of mathematics of his time. The two men, now friends, then begin a long scientific collaboration around geometric questions. They study together Euclid's *Elements*—a work of geometry (plane and in space) and theory of numbers and ratios—fundamental since Greek antiquity. Thus, Leonardo will brilliantly illustrate Pacioli's treatise with numerous

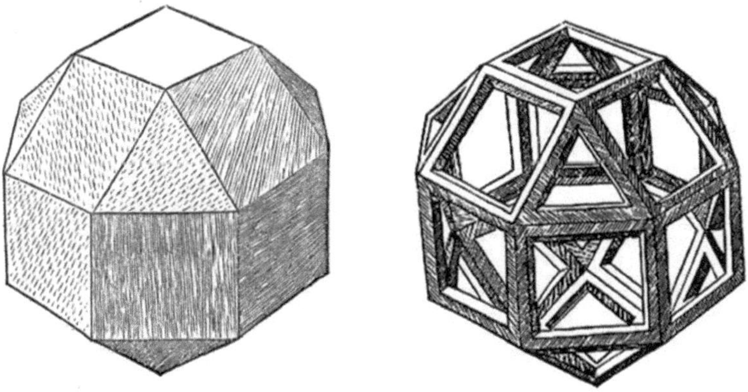

Fig. 8.1 The rhombicuboctahedron, a polyhedron with 26 faces (solid and empty) represented by Leonardo da Vinci in the Pacioli's *De divina proportione* [Divine proportion] (Paganino Paganini, Venise, 1509)

geometric representations, some of which are extremely complex (for example, Fig. 8.1).

In 1499, the city of Milan is occupied by the French and Leonardo leaves the city with his friend. He travels through Tuscany and Romagna as a military engineer for Cesare Borgia, his new patron. A few years later, around 1506, he arrives at the French court of Milan, governed by Charles d'Amboise who admires his work. All freedom is given to him for his own work: it is at this time that he begins, among other things, his "Mona Lisa" project which he will not complete until many years later. When in 1511, the French are driven out of Milan, Leonardo also leaves the city and takes refuge in various other cities. From 1513, he is welcomed in Rome by Pope Leo X who offers him his protection. It is there that he paints his self-portrait (Fig. 8.2). In 1515, the French troops, led by François I, invade Milan again. And from 1516, Leonardo comes to live in Amboise at the invitation of the young king. Even weakened by a paralysis of the right arm, he continues his creative work. In particular, he travels with François I to Romorantin to invent a new capital. He dies in Cloux in 1519.

Leonardo is a man of all arts and all sciences. He paints, he sculpts, he invents machines. All his works, whether realised or not, are based on a careful, meticulous observation of man, of fauna, of flora and of the properties of nature in general. The specialists thus describe Leonardo's scientific method: he starts from the observation of a phenomenon, tries to vary certain parameters, then he concludes, invents with the help of reasoning by analogy (that is to say by strict imitation). His scientific method proves powerful in several fields. He studies throughout his life, gradually enriching his personal library. His objects of observation and reflection are diverse such as birds and living beings in general, water and other fluids. In addition to painting, sculpture, architecture and urban planning, we can mention optics, perspective, and more generally geometry but also mechanical arts, geology, botany,

Fig. 8.2 Self-portrait of Leonardo da Vinci preserved at the Royal Library of Turin, © Web Gallery of Art

anatomy or even cardiology and embryology as scientific disciplines to which he devotes himself. Each time, the drawing occupies an important place.

Leonardo's Mathematics in the History of Mathematics

Let no one read me who is not a mathematician
["Non mi legga, chi non è matematico [...]", Windsor, IV, fol.163b]

This is how Leonardo expresses himself in a codex dedicated to anatomy and now preserved in the Royal Library of Windsor (England). This sentence is reminiscent of the one inscribed, in ancient Greece, on the front of Plato's Academy: "Let no one ignorant of geometry enter here". It is difficult to summarise Leonardo's mathematical work (Marinoni, 1982; Duvernoy, 2008). On the one hand, it is heterogeneous and scattered throughout his notes. On the other hand, while these autographs present many geometric figures, he himself is not very talkative about his work and provides very few explanations.

The Ratios of Magnitudes

Like many mathematicians, Leonardo works on the ratios of magnitudes (lengths, areas, volumes...). In other words, he does not necessarily seek to evaluate magnitudes themselves but rather the relationships they maintain with each other. This is how he represented the Vitruvian man, in reference to the architect and Roman engineer of the first century BC, from the ratios observed between all the members of the human body. Double, triple, quadruple are ratios as well as half, third, etc. Equality is expressed as a particular ratio. Leonardo has a clear desire to measure the world, in the broad sense.

To better understand the sheets that served as supports for educational activities (See Appendix 2), it is necessary to briefly present the ideas underlying Leonardo's constructions. As in other scientific fields, he will set up various variations of figures to observe the properties that are preserved, and to evaluate new magnitudes in relation to the old ones. Indeed, Leonardo is concerned with the problem, already posed in ancient Greece, of the squaring of the circle. This involves constructing—with a ruler and compass alone—a square that would be equal, in area, to a given circle. Leonardo is convinced that he has found various solutions to the problem (like many other mathematicians and enthusiasts after him).

However, by the end of the nineteenth century, the problem was proven to be impossible because the number π—the ratio between the circumference and the diameter of a circle—is transcendental (there are no non-zero polynomials with integer coefficients of which π would be a root). All the proposed constructions are only approximations, not exact. Therefore, it is up to Leonardo to work on areas. However, he is less concerned with evaluating the area of a figure than comparing it to the area of another similar or different figure. He specifically wants to relate the two areas to each other.

The Method of Areas by Transformation

From the many variations of geometric figures mainly built on the circle and the square, Leonardo thus states area ratios aiming at the equality in area of two figures. For this, he uses several distinct methods, two of which, easily understandable by pupils over 9 years old, are very useful.

The first consists of sliding a figure from its initial position to a final position and using the principle: *If equal parts of equal figures are removed, the remainder must be equal* ["Se di due cose equale se ne leva parte equale, el rimanente resta equale", *Codex Atlanticus*, fol. 1040v] taking up common notions from Euclid's *Elements*. Let's take an example. Consider Fig. 8.3: it consists of two equal discs (because of the same radius), one being obtained by sliding the other. By removing the central part (more grey) from both, the two parts A and B remain, which are of the same area.

The second method consists of a copy/paste. Let's take the example of Fig. 8.4. From the equilateral triangle, we cut a portion of a circle on one side to paste it on another side. We thus operate a transformation of triangle C towards the sickle-shaped figure D (in the shape of a scythe or sickle) which then has the same area.

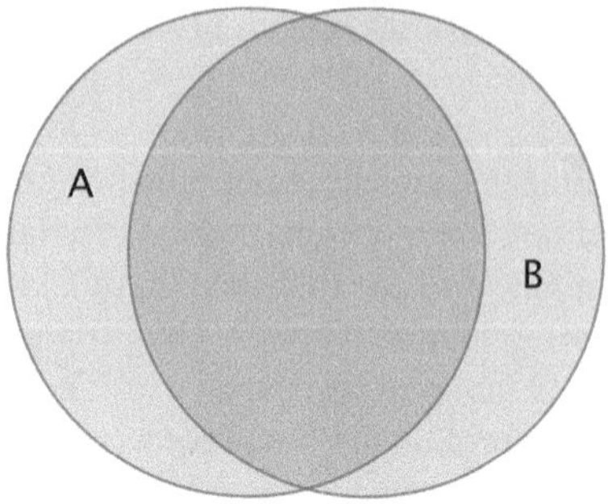

Fig. 8.3 Equality of areas by sliding and subtracting the common part

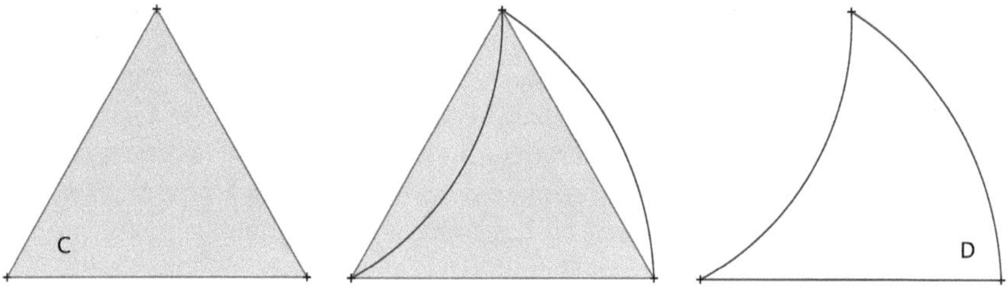

Fig. 8.4 Equality of areas by copy/paste

We wanted to explain the reasons for the geometric figures in the codex even if this is not what our pedagogical experiments will be based on (except perhaps in the envisaged extensions). But, without further ado, let's present the historical source that inspired us.

The Codex Atlanticus and Geometry

The *Codex Atlanticus* is one of the many notebooks and sketches of Leonardo (Piumati, 1904). It is actually a compilation entirely made posthumously by Leoni, an admirer of Leonardo and a sculptor at the court of Madrid. More than a thousand manuscript pages are thus gathered in 12 volumes. Today preserved in the Ambrosian library (in Milan), it owes its name to its large format (64.5 cm × 43.5 cm) reminiscent of atlases.

This codex perfectly illustrates the scientific life of Leonardo as it covers a very long period of his life: from 1478 to 1519, and illustrates all the versatility of the artist. Many drawings represent flying machines, war machines, musical instruments, various geometric figures; there are also preserved handwritten notes related to astronomy, geography, botany, architecture, anatomy, some autobiographical notes and other philosophical considerations. At this stage, it is important to understand that all these manuscript pages, extremely heterogeneous, are gathered because they are all the work of Leonardo. However, Leonardo would never have intended to gather them himself and publish them. These are drawings and personal notes that were not made or written to be communicated to anyone. We enter into the intimacy of Leonardo's workshop without explanation or concern for didactic presentation. Furthermore, Leonardo was left-handed and had the habit of writing from right to left and in mirror image. To understand him, the reader has to use a mirror. This makes Leonardo's writings even more difficult to access.

As far as this chapter is concerned, we present a study on the only folio 471verso (Fig. 8.5), dating from 1516, which shows the variations to which Leonardo indulges. We now need to detail our choice of figures to better meet our educational objectives.

Scientific and Pedagogical Aspects

Didactic Challenge of Reproducing Figures

Our work is mainly based on the didactic works of Marie-Jeanne Perrin-Glorian and her team (Perrin-Glorian, 2008; Perrin-Glorian & Mathé, 2023). They are the most recent English contributions, as syntheses of older works published in French (Offre et al., 2006; Keskessa et al., 2007; Perrin-Glorian et al., 2013; Duval et al., 2023). It is also necessary to mention the more general works on geometry made by Berthelot and Salin (1998, 2001).

> Doing geometry requires a specific look at the figures [...]:
>
> - One must be able to switch from a vision of juxtaposed figures to superimposed figures (and vice versa), by adding, erasing lines...
> - One must be able to see objects of several dimensions in the figures: surfaces, lines, points and relationships between them [...]. (Perrin-Glorian et al., 2013, p. 17)

In particular, understanding that a point is the intersection of two lines (not necessarily straight) is a challenge of basic education (8–10 years old). Therefore, it is necessary to set up rich problem-situations that gradually lead pupils to look for lines which, in turn, allow to construct points. For example, pupils must be able to see a vertex of a quadrilateral as the intersection of two sides and not just as a corner of the figure (the polygon being then considered only as a surface). Indeed, if a pupil remains in a "geometry of shapes", he will be unable to create new lines to obtain points (endpoints or intersections of lines) or to establish, for example, alignment relationships.

The figure to be reproduced is then considered as a support for researching the geometric relationships that can be established. It allows to experiment what can be traced and what is missing. The observation of this figure is an essential step to determine the necessary conditions of the problem.

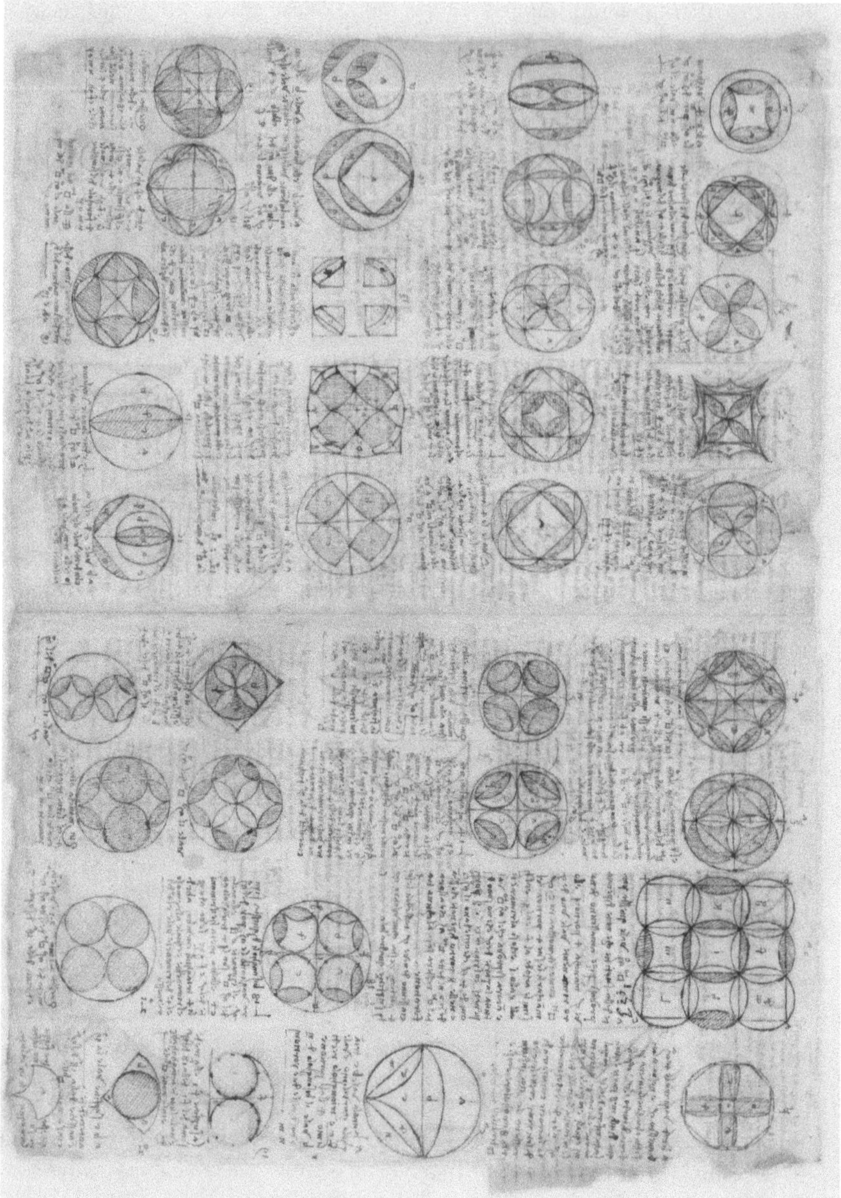

Fig. 8.5 Folio 471v of the *Codex Atlanticus* showing a series of geometric figures

Pupils are therefore led to create points by intersection of two lines and, in turn, the points themselves define lines. Two points characterize a segment but also a circle (the centre and a point on the circle). Two points can also determine a half-line or a straight line.

Throughout our work, we do not impose any measurement and therefore, *a fortiori*, we do not ask for an "identical reproduction". Indeed, we thus focus on the geometric properties of figures without taking into account the size of the elements. The figures thus reproduced are all reductions or enlargements of the original figure. We do not exclude the use of the graduated ruler to determine the middle of a segment; the use of the compass alone (construction of the bisector) is here too difficult for the majority of 9–10 year old pupils. The didactic challenges related to measurements can be extended in the field "magnitudes and measures" with work on areas and perimeters (Section "Sizes and Measurements: Area and Perimeter for 10–11 Year Olds").

The Choice of Figures to Reproduce and Analysis A Priori

If the exercise of reproducing figures is often quite standard for pupils aged 6–14, the teacher's choice is not always easy, often due to a lack of different and geometrically rich figures for use in the classroom. To make our choices from Leonardo's manuscript, we had lengthy discussions in our group. In relation to the didactic issues seen above, we discarded all the figures that could not be easily constructed by identifying alignments and only intersections of straight lines, segments or arcs of a circle. Of course, the whole nuance rests on the adverb "easily". In particular, the teacher must always question the possibility offered to pupils to find the characteristic points of the figure. Therefore, one should not believe that the more sober a geometric figure is, the easier it is to construct. It's a trap! The simplest figures to the eye are often the most difficult to execute.

Let's take the following Fig. 8.6, composed of a large circle and four inner circles tangent to each other and to the large circle.

The aim here is to highlight the relationships between the circles, and therefore the centres and the respective radii or diameters. Several constructions are possible. We detail one from a square whose four vertices are the centres of the four small tangent circles (Fig. 8.7).

If a pupil had to reproduce this figure, it would therefore be a difficult (if not impossible) task as they would have to precisely imagine points (the centres of the

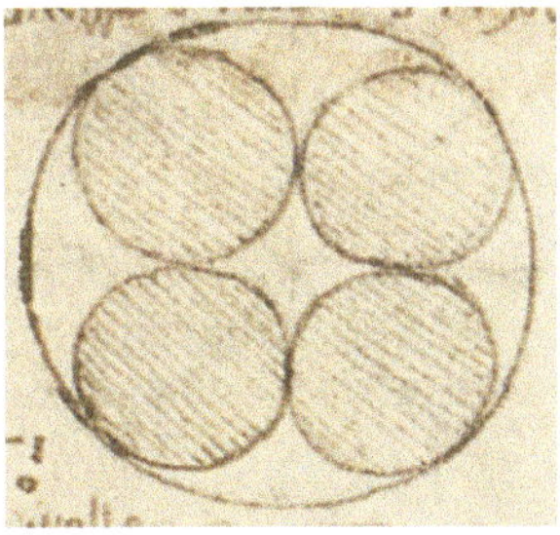

Fig. 8.6 Example of a sober yet complex figure (fol. 471v, top left)

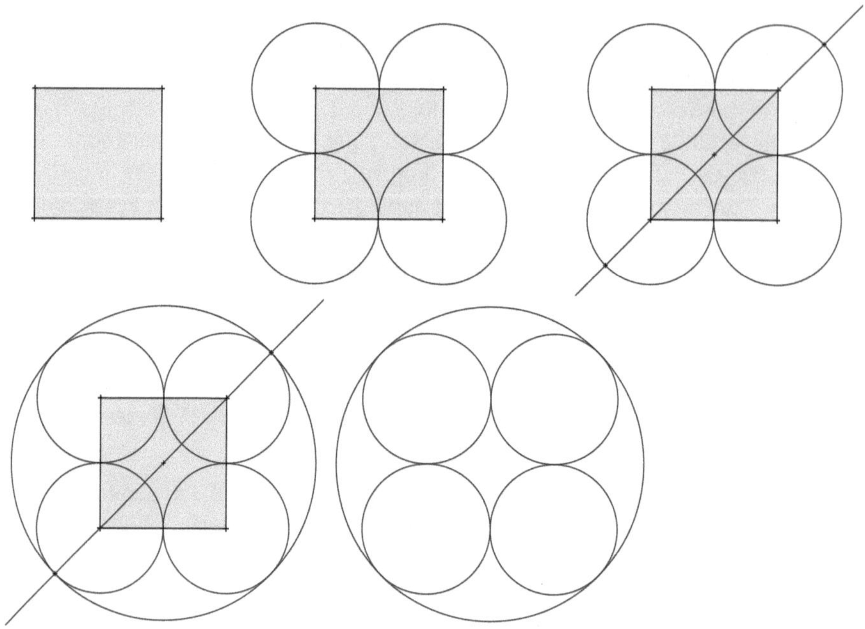

Fig. 8.7 Possible construction of Fig. 8.6

circles corresponding here to the vertices of the square, itself absent from the initial figure), and alignments (those, for example, determined by the line carried by a diagonal of the square). Therefore, great care must be taken in the choice of figures and not just be guided by the aesthetics or apparent simplicity of the figure.

We first imagined a session with five different figures but the collective synthesis proved impossible to implement. We therefore definitively chose to limit ourselves to two figures, close to each other (Figs. 8.8 and 8.9), in order to be able to make collective assessments both to help pupils during the execution and for a final synthesis highlighting the vocabulary used, the definitions and other properties implemented. Moreover, this choice allows for differentiation within the class. Indeed, even if the tasks to be performed will be the same for the two figures, they do not require the same reflection. To be reproduced, the second figure (Fig. 8.9) requires the pupil to necessarily reveal at least one element that is not in Leonardo's.

For these two figures, one must think about the chronology of the construction. If one starts with the encompassing figure, that is the large circle, the construction is then difficult. Indeed, it is necessary to construct two perpendicular diameters,

- either to actually construct the large square of the first figure (Fig. 8.8) absent from the second figure (Fig. 8.9); a quadrilateral being a square if and only if its diagonals are of the same length and intersect perpendicularly in their middle,

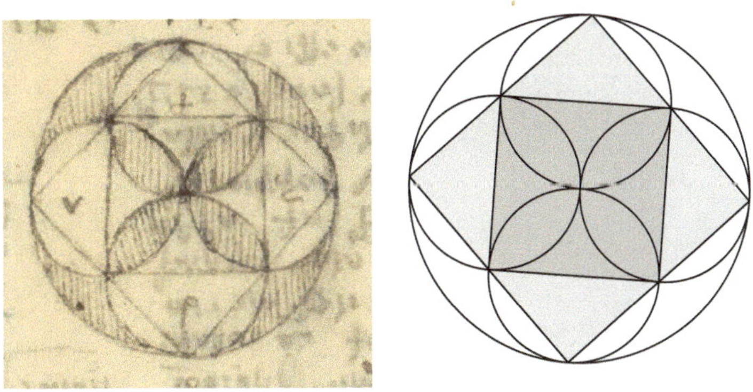

Fig. 8.8 First figure chosen for implementation in the classes (fol. 471v, in the centre)

Fig. 8.9 Second figure chosen for implementation in the classes (fol. 471v, in the centre)

- or to determine, in the second figure, the points of tangency of the four small circles with the large one. The large square of Fig. 8.8 can then be seen as a help because it is visible and therefore must be constructed.

If, on the other hand, the construction begins with that of the small square, Fig. 8.8 is then constructed without too much difficulty. For Fig. 8.9, the task remains complex because it is still necessary to consider the points of tangency (or contact) of the large circle with the four small circles.

The construction of the two figures reveals the middle of a segment. As we have already mentioned, for pupils aged 9–11, this point is generally determined using the graduated ruler; too few pupils use an expert procedure with the compass. If one wants to avoid constructing the middles, one must either start from the small square and draw perpendiculars to the sides passing through its centre, or start from one of the four small circles, draw perpendiculars and report the radius. We have chosen not to impose any starting point on the pupils to make their analysis more fruitful.

We give below two possible constructions (Figs. 8.10 and 8.11) for Fig. 8.8. Those for Fig. 8.9 are simply deduced from them.

Let's now detail our experiment in class.

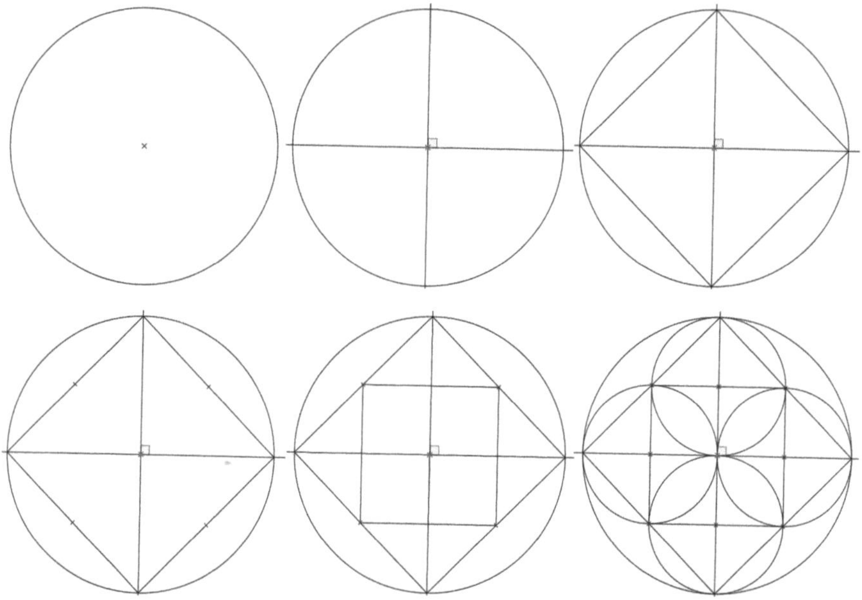

Fig. 8.10 A possible first construction of Fig. 8.8

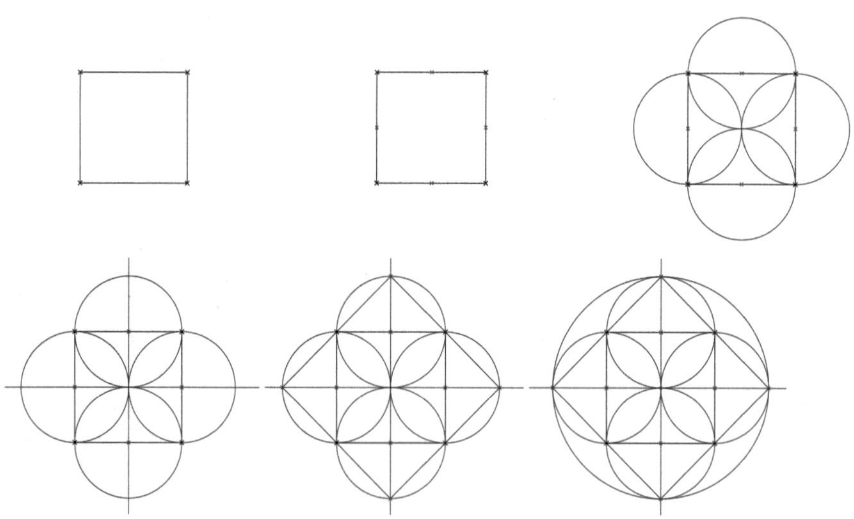

Fig. 8.11 A possible second construction of Fig. 8.8

Implementation in the Classroom

The activities proposed and analysed below have been tested in six classes: three classes of 10–11 year olds, in two classes of 9–10 year olds and in a class of 8–9 year olds.

The work was systematically carried out in the whole class: two work sessions (1 h) in small groups and a collective session, shorter, for the synthesis (30 minutes).

Reasoned Observation or "With Intention"

With the aim of "casting a specific gaze" on the figures to reproduce, we thought it was wise to create a freehand figure to appropriate it. But we realised that freehand drawing was more about mimicry than a reasoned analysis of the figure that would have allowed to highlight the properties of it and its characteristic points.

Often, pupils identify the elements of the figure (the geometric shapes like, here, the squares and the circles) but do not establish the relationships between the sub-figures in order to enrich the model figure with new traces that will be used for its construction. For example, pupils did not notice that the centres of the inner circles are the midpoints of the four sides of the small square. They will then be unable to reproduce the figure using the instruments.

We preferred to set up a time for observation (individual then collective) of the figure to be reproduced, encouraging pupils to identify, on it, the constituent elements (Fig. 8.12) and to think about a possible order for the construction steps (Figs. 8.13 and 8.14). It is absolutely necessary to let the pupils (about 10 minutes) appropriate the complex figure and the elementary figures that make it up.

A group discussion is then gradually set up where everyone will detail the way they observed the figure (up to 30 minutes). Indeed, it is essential to determine an order of construction in relation to the instruments (ruler, set square, compass) to be used. If we start with the circle, it is more difficult to draw the square: it must be obtained by its diagonals which are two diameters of the circle not drawn on the

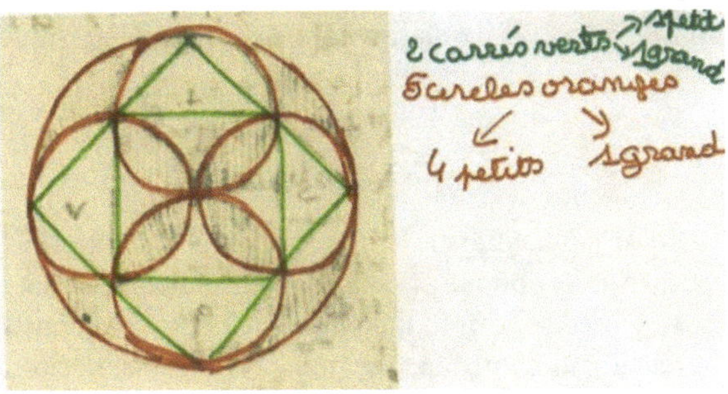

Fig. 8.12 Identification of the figures present: 2 green squares (1 small, 1 large), 5 orange circles (4 small, 1 large)

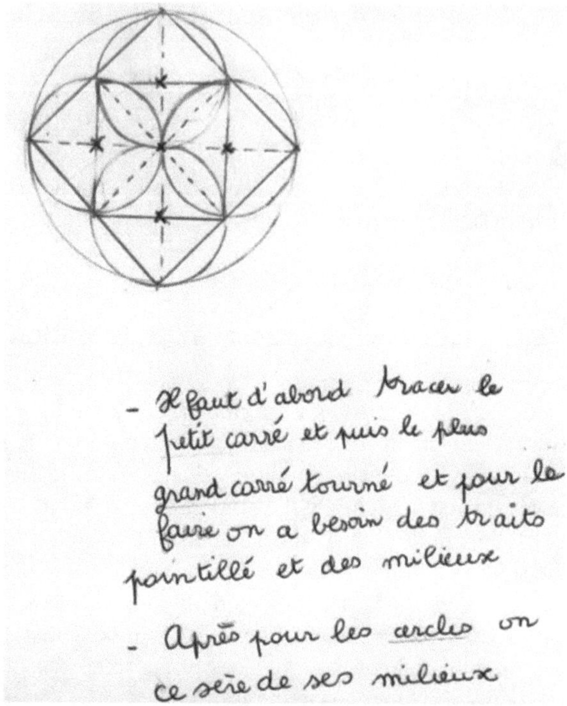

Fig. 8.13 Steps according to the pupils (English translation from the French): First you have to draw the small square and then the larger rotated square. And to do this, we need the dotted lines and the midpoints. Then, for the circles, we use these midpoints

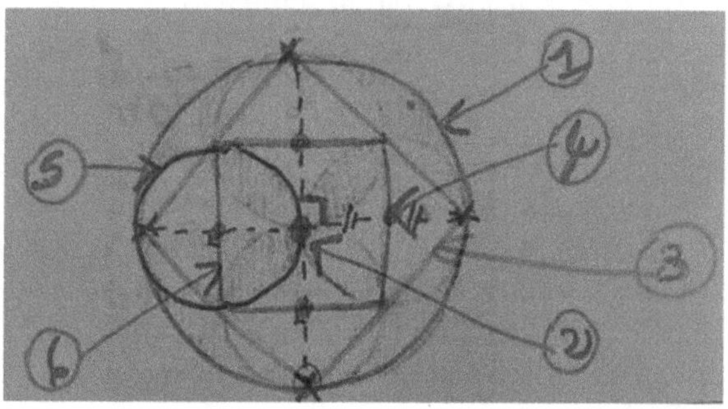

Fig. 8.14 Numbered steps on figure: (1) The large circle; (2) Two perpendicular diameters (identified with coding); (3) The large square; (4) The midpoints of the radii corresponding to the centres of the small circles (identified with coding); (5) The four small circles; (6) The small inner square

starting figure. If we start with the square, we must identify the centre of the square as the centre of the circle.

Two main approaches stand out, starting from the inside towards the outside or vice versa.

Some remarks:

No group having chosen to start its construction with the large square (probably because of its oblique orientation), one could consider a situation of restoring figures starting precisely from this one. The geometric constraints would then be set from the square, its centre and the middles of its sides (circumscribed circle and inscribed square).

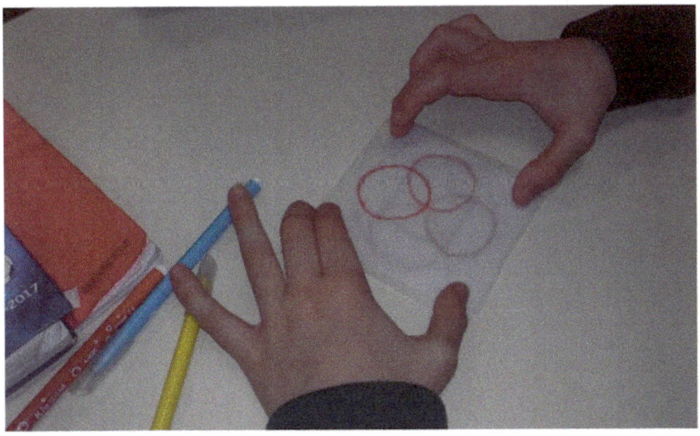

Fig. 8.15 Use of overlays (Photo: Marc Moyon)

The figures being assemblies of simple figures obtained by superposition, for the youngest pupils (8–9 years old), it may be wise to use overlays (Fig. 8.15) to decompose the figure. This avoids, in particular, that the vision of the petals in the square is predominant compared to those of the circles: a petal can only be the result of the intersection of two circles.

The Reproduction of the Figure

The pupils will validate their previously described steps by realising the figures. They then embark, for the most part, on a trial/error approach. Indeed, almost all have the natural attitude to start with the encompassing figure: here, the large circle (Fig. 8.10). However, for the youngest pupils (8–9 years old), the construction of the square inscribed in a circle is difficult (if not impossible) because it is linked to the fact that the pupils do not have the idea to try to make the visible figures (here, the sides of the square) appear by tracing auxiliary lines both invisible on the figure and much more unusual than the sides (here, the diagonals of the square—a novelty for most pupils—while the square is encountered from the earliest childhood, even before 6 years old as a geometric shape). The pupils then end up thinking about the construction from the small square (Fig. 8.10).

The older pupils (10–11 years old) have worked and learned the geometric property of the diagonals of the square (perpendicular, of the same length and intersecting in their middle); thinking about the diagonals is then less difficult. Even if it is not a specific objective for the pupils (8–11 years old), some of them are even capable, if they wish, to justify that they are indeed drawing a square.

The main problem then lies in the construction of a square on white paper (Fig. 8.16), the pupils using the ruler and the set square. The errors in constructing the right angle or two perpendicular lines add up at each corner of the square, the determination of the lengths of the sides with the ruler is barely more precise. To

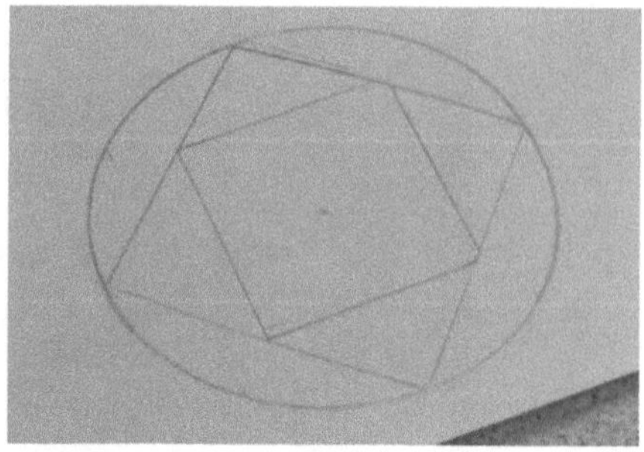

Fig. 8.16 The difficulty of drawing a square on white paper (Photo: Marc Moyon)

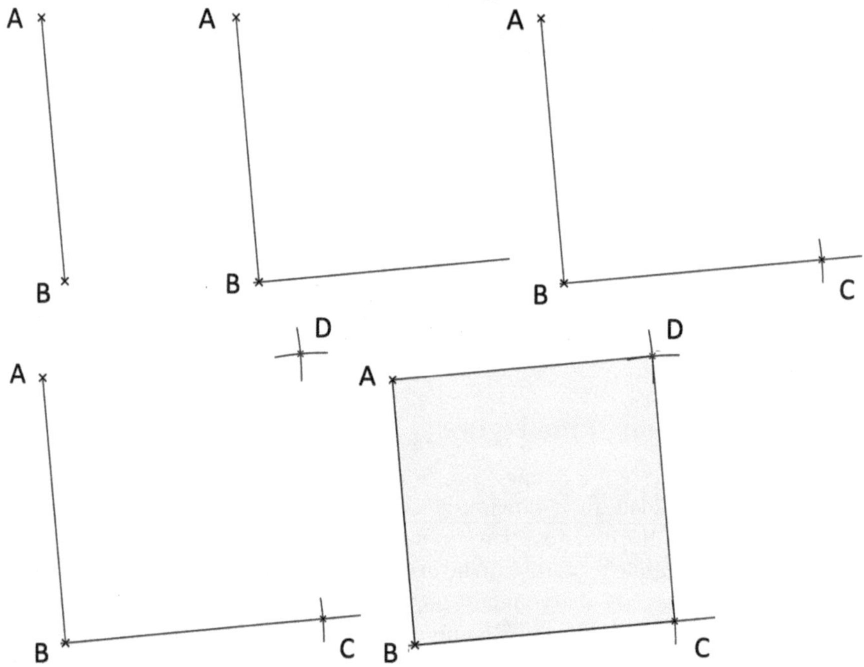

Fig. 8.17 Construction of square $ABCD$ from a right angle (using the square at vertex B) and three length transfers (with the compass for vertices C and D)

overcome this obstacle, it may be suggested to resort to the use of graph paper (which can also be used as a differentiation tool).

The use of appropriate geometric instruments is an explicit mathematical aim for pupils and must evolve significantly between 8 and 11 years old. It is necessary to go beyond the construction with the ruler and the set square alone (here, four right angles and four equal sides). For some 9–10 year old pupils and for all those of 10–11 years old, a variant to the ruler-set square construction is then considered: from a vertex from which two perpendicular sides of the same length are drawn with the set square and compass, the fourth vertex is constructed with the compass alone (Fig. 8.17).

Finally, one can also gradually impose the sole use of the ruler and compass (with the help of the bisector).

Provided they have reasoned enough about the figure to be reproduced, the pupils have the different construction steps well in mind, these having been validated by the group. Each pupil then carries out his or her individual construction (Figs. 8.10 and 8.11).

Drafting a Construction Program

Drafting a construction program is only a goal for the oldest pupils (10–11 year olds). Therefore, it is out of the question for us to ask for this drafting as the first task.

For 8–9 year old pupils, the drafting of the construction program is done orally, with the whole class to synthesise the activity. This is an opportunity to revisit the different elements of the figure with the appropriate vocabulary and the steps of the construction. The role of the teacher is then to help the pupils structure the text.

For 10–11 year old pupils, the work has to be very progressive. The pupils are first invited to write a construction narrative, that is, a text that is both narrative and explanatory, covering the different geometric figures and the steps of construction. The texts are often more descriptive than operational, meaning they are not sufficient to draw the figure. However, if passed on to someone who has not seen the figure, the construction program should allow them to reconstruct it. Here, at first, we are far from it.

For example, about Fig. 8.9: "It is composed of 5 circles and 1 square. I started by making the square. Then I built the four small circles that form a small rosette. Then I made the large circle all around the figure."

These texts cannot be validated by the other pupils in the group. As a direct consequence, many of them are led to name certain points of the figure (Fig. 8.18) in order to structure their writing and avoid any misunderstanding in the many points to be placed (centre of the circle or square, middles of the sides, corners of the squares in particular). This is an attitude to encourage. Indeed, to write a suitable construction program (correct and succinct), it is often necessary to name the points and know that one can designate a length, a segment or a line by two characteristic points (for example, Fig. 8.19).

In small groups, the pupils must then agree on a common text formalised as a construction program, a type of text that has already been encountered at school in previous years (before 8 years old). As in the construction programs already worked on in class, the verbs to draw, to place or to name are favoured. It is necessary to clearly describe the geometric figures by their characteristic elements and not by their shape and/or size. For example, a circle is characterised by a point (its centre) and a length (its radius or diameter), or by two points (its centre and a point on its circumference). The construction program can no longer feature a "small" or a "large" square or circle. For the circle, this proves difficult as the pupils have not

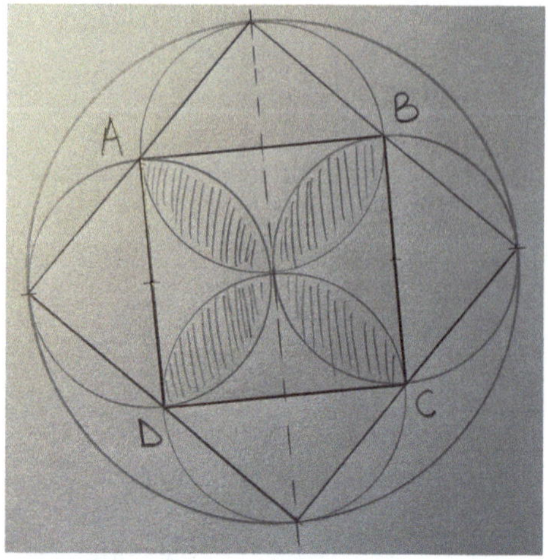

Fig. 8.18 Indication of points on Fig. 8.8 made in relation to the construction program

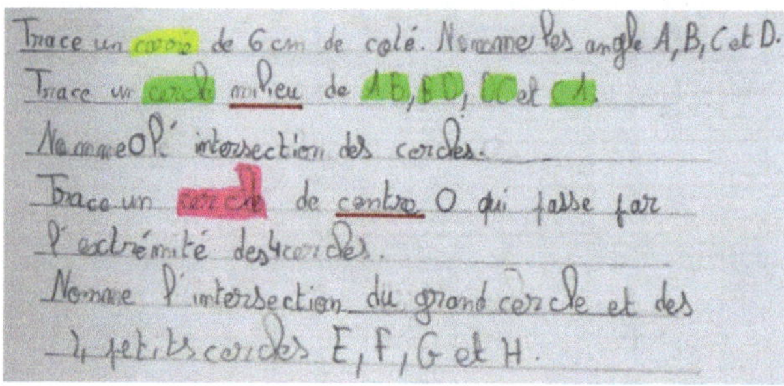

Fig. 8.19 Example of a construction program for Fig. 8.8 drafted by a group of 10–11 year old pupils. (English translation) "Draw a square with sides of 6 cm. Name the angles A, B, C and D. Draw a circle in the middle of AB, BD, DC, CA. Name the intersection of the circles O. Draw a circle with centre O that passes through the ends of the 4 circles. Name the intersection of the large circle and the 4 small circles E, F, G and H"

yet got into the habit of giving a name to this figure (which, unlike polygons, has no vertices to designate it).

Here is an example of a collectively constructed construction program, relating to Fig. 8.14:

> Draw a circle with centre O. Draw $[EF]$ and $[GH]$ two perpendicular diameters. Draw the quadrilateral $EGFH$. Place the points M, N, L and K, respective midpoints of the sides $[EH]$, $[HF]$, $[FG]$ and $[GE]$. Draw the quadrilateral $MNLK$. Construct the circles with respective diameters $[MN]$, $[NL]$, $[LK]$ and $[KM]$.

Once the program is written and accepted by all, it is presented to the whole class in the form of a poster. All the programs thus obtained can be tested by pupils from other classes (of the same level or lower) to be validated or invalidated.

For all pupils (8–11 years old), several other devices can be used to differentiate the work. In particular, the teacher can ask some pupils to reorder a construction program whose steps have been previously cut out and disordered. We can also, always in accordance with official instructions, ask some pupils to complete a construction program already written by inserting words, noun groups or entire sentences that are missing.

Ways to Go Further

The different parts that follow are just ideas for possible extensions. We cannot detail them further here. We developed elsewhere another activity with the *Codex Atlanticus* (fol. 1040 (recto)) with a square and triangles inscribed within a circle. It is developed and analysed in (Moyon & Plantevin, 2025).

Elsewhere, in the Mathematics Program

Sizes and Measurements: Area and Perimeter for 10–11 Year Olds
To illustrate our work on sizes and measurements, we have chosen another figure from the *Codex Atlanticus* (Fig. 8.20).

This figure is particularly interesting to run for the calculation of its perimeter and its area. Indeed, for example, from a photocopy of this figure (enlarged on an A4 sheet), the oldest pupils (10–11 year olds) can develop different strategies to measure its perimeter and its area, from the most basic (measurements, cut-outs, assemblies, colouring of surfaces or lines) to the most expert (sequence of calculations using the fewest possible steps for example).

A pooling of productions with comparison of approaches and validation or not of the results provides an opportunity to take stock at the end of the "area and

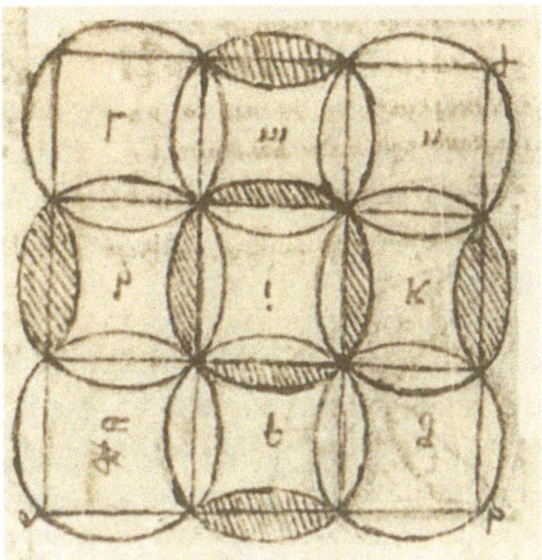

Fig. 8.20 Geometric figure by Leonardo da Vinci, *Codex Atlanticus*, fol. 471v

perimeter" chapter. From our observation in class during an experiment, the main errors identified come from:

- confusion between area and perimeter,
- difficulty working on portions of circle and disc,
- organisation of calculations.

With Axial Symmetry

The historical document (Leonardo's sheet: Fig. 8.5) can still be used during activities concerning axial symmetry: search for axes of symmetry of a figure, search for the pattern to reproduce by symmetry to find the figure with the possibility of validation using a geometry software.

Issues in History and Geography

Working on Leonardo da Vinci requires going through the presentation of the historical context, which is very rich.

The teaching of history shows that historical landmarks "allow us to understand that today's world and contemporary society are the heirs of long processes, of ruptures, of choices made by the women and men of the past" (taken from the French teaching program (Ministère de l'Éducation nationale, 2015, 173)). It can be a question of working on "The time of kings" which allows to work on François I as a protector of Arts and Letters during the Renaissance, his role as a patron through the specific example of Leonardo da Vinci.

Ultimately, it is Leonardo da Vinci himself who is the link, or more precisely the "hyphen" between Mathematics, History and Geography. The reading of his biography (Section "Leonardo da Vinci (1452–1519): Some Biographical Landmarks") is proof of this: his research fields are so varied that all school subjects can be approached from Leonardo da Vinci. History and geography are no exceptions. The common thread for teaching history could be, for example, "Humans represent the world", it can lead to (at least) two concrete achievements. In History, pupils construct a mental map (Fig. 8.21) on this theme from the lessons and documents proposed by the teacher to continue and open up to the Middle Ages. In Geography, pupils are led to imagine (drawing and legend) the city of tomorrow as Leonardo could have done for Romorantin, a French city. We can work on the two videos on this subject, which can be shown to pupils (in French): "Romorantin, the ideal city of Leonardo da Vinci" (http://culturebox.francetvinfo.fr/patrimoine/romorantin-la-cite-ideale-de-leonard-de-vinci-exposee-au-musee-de-sologne-39979, accessed on June 30, 2025) and "Romorantin, the forgotten project of Leonardo da Vinci" (http://www.romorantin-leprojetoubliedeleonarddevinci.fr/, accessed on June 30, 2025).

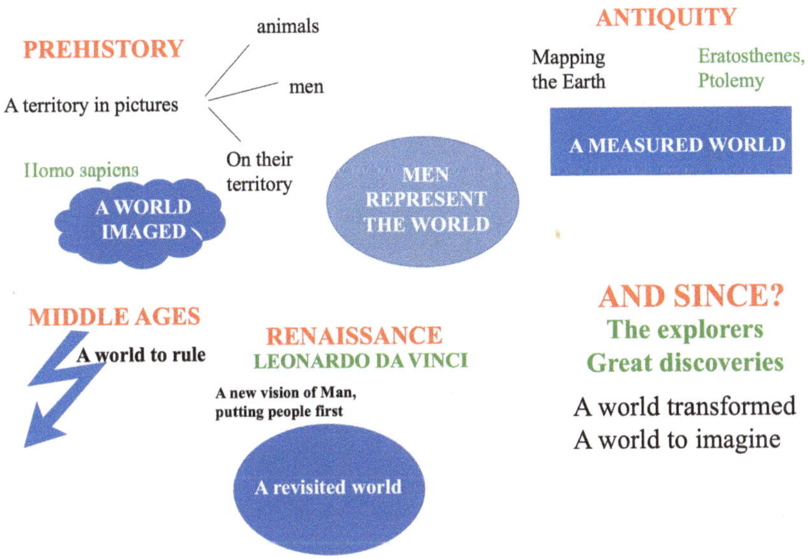

Fig. 8.21 Structure of the mind map created by the 10–11 year old pupils

For this, the teacher can lead his pupils—occasionally after lessons that lend themselves to it or during a special session—to become aware that men have, at all times, represented the world: projection of maps of the known world with the examples of Eratosthenes in the Third century BC, Ptolemy in the second century, al-Idrīsī in the twelfth for the most known representations... They explained it using the knowledge of the time and some, like Leonardo, even imagined how to improve their daily life.

The teacher can even place himself in a long historical perspective by introducing the common thread through cave paintings that give us a representation of the world, or rather of the territory of Paleolithic and Neolithic men. The history and geography teacher can then open up to a geography theme around "Living in a metropolis". Indeed, Leonardo da Vinci is a precursor in the field of urban planning. His project of an ideal city in Romorantin can be explained to pupils and allows to chain on Geography and to evoke our way of representing the world today with modern tools (*Google Earth*, for example). It still allows to open up to current projects of the city of the future. Pupils are finally invited to carry out their project of the city of tomorrow in the form of a drawing to be accompanied by a legend, or even explanations (Fig. 8.22).

Thus, History and Geography complement each other perfectly to provide a different perspective on this emblematic character of the Renaissance. The notion of "common heritage" takes on its full meaning here. The documents presented by the teachers can reveal the diffusion of knowledge (between the Christian world and the Muslim world on the first world maps: map of Ptolemy taken up by the Arab geographer al-Idrīsī for example) and man's questioning of the world (with the Greek scholars of Antiquity in particular).

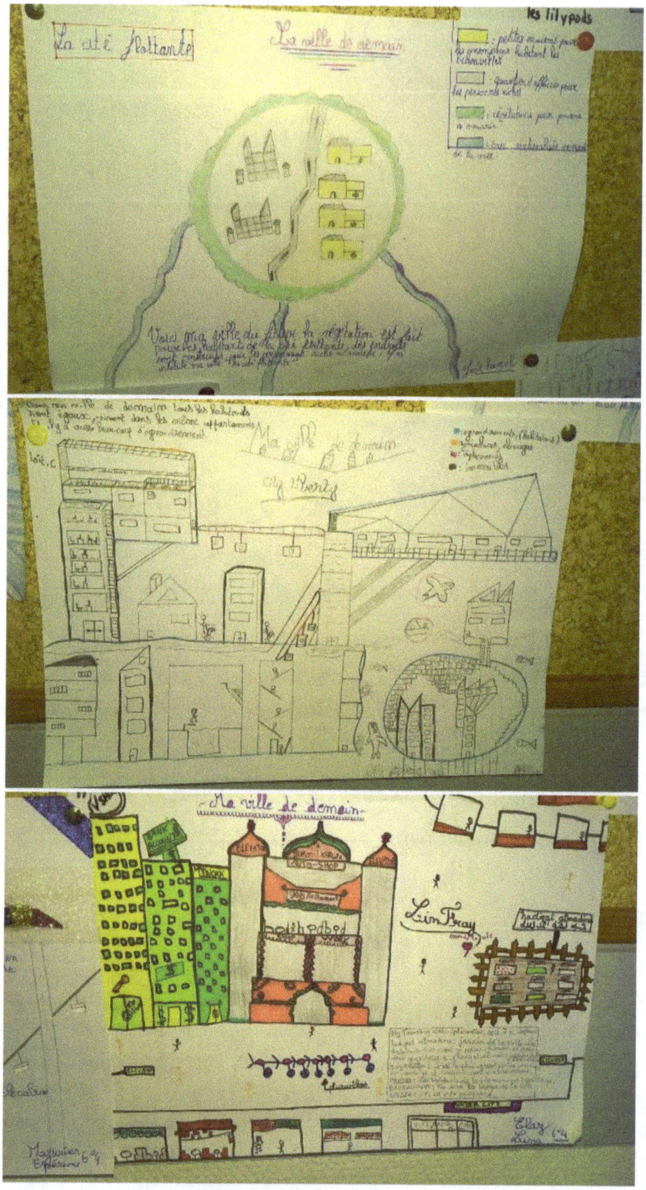

Fig. 8.22 Ideal cities invented by the older pupils (10–11 years old) (Photo: Marc Moyon)

And with Dynamic Geometry Software?

It is possible to ask pupils to reproduce one or more of these figures (among those of fol. 471v. in colour) with geometry software (like *GeoGebra*).

This work can be proposed either after having made the figure with the instruments (in this case the pupils repeat the steps of their construction program), or before, or at the same time. We can imagine two groups of pupils working one on paper, the other on screen, each group having to explain its construction to the other at the end. The objectives are then different.

In the last two cases, the use in the menu of the regular polygon to draw the square can lift the difficulty of its construction.

For other figures, on the contrary, the constraint of constructing the square with a compass can be imposed, hence proposals for differentiated work according to the pupils in the context of the implementation of personalised support sessions for example.

Acknowledgments This chapter was written by Marc Moyon with the help of Chantal Fourest and Valérie Rosier-David from the Secondary School d'Arsonval (Brive, France), David Somdecoste from the Primary School Louis-Pons (Brive), Jérome Dufour (†) from the Secondary School Cabanis (Brive) and Véronique Lefranc from the Primary School Roger-Gouffault (Brive). All these people worked together within the research and reflection team "School/college liaison and history of mathematics" of the IREM of Limoges (France).

Appendix 1: Concepts and Skills Addressed in This Chapter

Concepts Covered

- Recognise, name, describe, reproduce, represent, construct geometric figures
  - Study of simple or complex figures (assemblies of simple figures)
  - Characterisation of plane figures (square, circle, triangles)
  - Use of appropriate vocabulary to name and describe
  - Reproduction of simple or complex figures
  - Drafting and execution of a construction program
- Recognise and use some geometric relationships
  - Perpendicularity, parallelism (construction of parallel lines, link with the property relating parallel and perpendicular lines)
  - Alignment, belonging
  - Equality of lengths

Skills Developed

- **Search and research skills**: Extract and organise the information necessary to reproduce figures from Leonardo da Vinci's drawings. Observe Leonardo's complex figures and manipulate the appropriate instruments. Test, try several drawings by varying the order of the construction steps.
- **Reasoning**: Gradually move from perception to control by instruments (non-graduated ruler, square, compass) to initiate reasoning based solely on properties of figures (circle and square) and on relationships between objects (intersection points, middle, centre). Progress collectively in an investigation knowing how to take into account the point of view of others (group work).
- **Mathematical communication skills**: Gradually use an adequate vocabulary and/or adapted notations to describe, orally and in writing, a complex geometric figure.
- **Mathematical representation ability**: Analyse a plane figure from different aspects (its outline, lines and points). Use elements of coding a plane figure to construct it and write its construction program.

Appendix 2: Worksheets for Pupils

Available at
https://www.univ-irem.fr/chapitre-8-la-geometrie-des-carnets-de-leonard-de-vinci

DISCOVERING THE GEOMETRY OF *Leonardo da Vinci*

If you don't know *Leonardo da Vinci*, here is some information that accompanies his self-portrait...

Leonardo was born in 1452 in Vinci, a small town in Italy, and he died in 1519 in Amboise (France). He worked for very rich and important princes, like Ludovico Sforza in Milan and King François, the first, in France. Leonardo is one of the most famous people from the Renaissance because of his art (paintings and sculptures), his drawings of the human body and plants, and his many inventions, although we don't know if he actually built them. Leonardo can also be called a mathematician. He even worked with another famous mathematician, Luca Pacioli, and drew geometric shapes for him. Unfortunately, Leonardo's writings are not easy to read because he wrote from right to left, in mirror writing!

Royal Library, Turin

By visiting the great Ambrosian Library in Milan, Italy, you can find an old and very important book of Leonardo da Vinci's work. This book is called the *Codex Atlanticus*, named after its large size ($64.5 \times 43.5cm$) – it's twice as big as the largest notebook a student uses today

On the back of page 471, you can find many geometric drawings that the Italian engineer and artist made himself. Here is one of them :

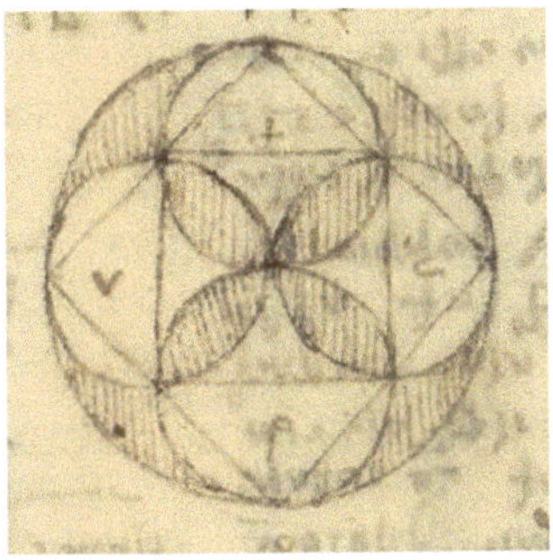

From « *De ludo geometrico* » : *la matematica e la geometria di Leonardo*, De Agostini, 2013.

Study of the Figure
Look at this figure very carefully. To do this, you can identify the basic shapes and name them. If needed, you can trace their outlines with a color.

Reproducing the Figure
Build it using the right tools.

Writing a Construction Text
Describe the different steps of this construction.

References

Ackermann, J. S. (1998). Leonardo Da Vinci: Art in science. *Daedalus*, *127*(1), 207–224.

Berthelot, R., & Salin, M.-H. (1998). The role of pupils' spatial knowledge in the elementary teaching of geometry. In C. Mammana & V. Villani (Eds.), *Perspectives on the Teaching Geometry for the 21 Century, an ICMI Study* (pp. 71–77). Springer.

Berthelot, R., & Salin, M.-H. (2001). L'enseignement de la géométrie au début du collège. comment concevoir le passage de la géométrie du constat à la géométrie déductive? *Petit x*, *56*, 5–34. https://irem.univ-grenoble-alpes.fr/medias/fichier/56x1_1561101962282-pdf

Denizeau, G. (2016). *Léonard de Vinci: le génie visionnaire*. Larousse.

Duval, R., Godin, M., & Perrin-Glorian, M.-J. (2023). Reproduction de figures à l'école élémentaire. In C. Castela & C. Houdement (Eds.), *Actes du séminaire national de didactique des mathématiques* (pp. 5–89). ARDM, IREM de Paris 7.

Duvernoy, S. (2008). Leonardo and theoretical mathematics. *Nexus Network Journal*, *10*(1), 39–49.

Isaacson, W. (2017). *Leonardo da Vinci*. Simon & Schuster.

Keskessa, B., Perrin-Glorian, M.-J., & Delplace, J.-R. (2007). Géométrie plane et figures au cycle 3. Une démarche pour élaborer des situations visant à favoriser une mobilité du regard sur des figures de géométrie. *Grand N*, *79*, 33–60. https://irem.univ-grenoble-alpes.fr/medias/fichier/79n3_1554797089105-pdf

Marinoni, A. (1982). *La matemática di Leonardo da Vinci: una nuova immagine dell'artista scienziato*. Arcadia Edizion.

Ministère de l'Éducation nationale. (2015). Programme d'enseignement du cycle des approfondissements (cycle 4). Journal officiel du 24/11/15. https://www.education.gouv.fr/pid285/bulletin_officiel.html?pid_bo=33400

Moyon, M., & Plantevin, F. (2025). Teaching mathematics through historically-based activities: Two educational experiments. *HISTEMAT. Revista De História Da Educação Matemática*, *11*. https://www.histemat.com.br/index.php/HISTEMAT/article/view/758/576

Offre, B., Perrin-Glorian, M.-J., & Verbaere, O. (2006). Usage des instruments et des propriétés géométriques en fin de cm2. *Petit x*, *72*, 6–39.

Perrin-Glorian, M.-J. (2008). Teach plane geometry in coherence from 6 to 15 years. *Journal of Mathematics, Teaching and Culture–REMATEC*, *48*. https://www.rematec.net.br/index.php/rematec/article/view/588/541

Perrin-Glorian, M.-J. & Mathé, A.-C. (2023). The geometry of tracing, a possible link between geometric drawing and Euclid's geometry? In C. Guille-Biel Winder & T. Assude (Eds.), *Articulations between tangible space, graphical space and geometrical space: Resources, practices and training* (pp. 3–34). Wiley-ISTE.

Perrin-Glorian, M.-J., Mathé, A.-C., & Leclercq, R. (2013). Comment peut-on penser la continuité de l'enseignement de la géométrie de 6 à 15 ans ? Le jeu sur les supports et les instruments. *Repères-IREM*, *90*, 5–41. https://irem.univ-grenoble-alpes.fr/medias/fichier/iwr13003_1704892231717-pdf

Piumati, G. (1894–1904). *Il Codice Atlantico di Leonardo da Vinci, in the Ambrosiana Library in Milan*. Hoepli.

Marc Moyon is Professor of History of Mathematics at the University of Limoges and a member of the XLIM Institute (UMR 7252, CNRS). He is also *Professor Catedrático Convidado* at the University of Aveiro (Portugal) within CIDMA (Center for Research and Development in Mathematics and Applications). His research focuses on medieval mathematics and the intercultural transmission of scientific knowledge, especially through Arabic-Latin translations, as well as the history of mathematics education in the nineteenth and twentieth centuries. He actively contributes to introducing historical perspectives into mathematics education and teachers training, notably within the national French IREM network (Institute of Research on Mathematics Education), and the International Study Group on the Relations between the History and Pedagogy of Mathematics (HPM) affiliated to ICMI.

Chapter 9
Protect Yourself with Mathematics: The Geometry of Fortifications

Frédéric Métin

> *Practice is as blind without Theory as Theory is crippled without Practice.*
>
> Jean Errard (1600, preface)

Abstract "The geometry of fortifications" was the mathematical part of an activity carried out in two classes of a Dijon primary school, as part of the *War and Science* project aimed at rediscovering the First World War from a scientific point of view. Among various activities in physics (projectile trajectories, war machines, etc.), natural sciences, and even ethics, the problem of fortress defence was addressed through geometric constructions we found in seventeenth century books, particularly the fortification treatise by Jean Errard de Bar-le-Duc, chief engineer of King of France Henry IV. In this treatise, all constructions are justified by calculations of angles and distances determined by the range of the weapons used. Thus, the pupils were guided towards understanding these constraints, as well as their modelling by mathematics and the demonstration of the adequacy of the shapes created by the military engineers to these constraints. A final creative activity invited pupils to give a personal signification to the fortress as their own imaginary city.

Keywords Military architecture · History of geometry · Geometry of war · Mathematical drawing · Military engineers controversies · Original documents in the classroom

The military architecture of the seventeenth century offers many advantages for the acquisition and exercise of geometric skills, whether it's about analysing or

---

F. Métin (✉)
IREM de Dijon, Université Bourgogne-Europe, Dijon, France
e-mail: frederic.metin01@u-bourgogne.fr

M. Moyon et al. (eds.), *Teaching Mathematics Through Historically-Based Activities*, History of Mathematics Education,
https://doi.org/10.1007/978-3-032-16990-7_9

drawing figures. Indeed, the shapes of citadels are both complex and regular enough to be visually apprehended in their entirety, from period plans or recent aerial photographs. Present in the immediate environment of many cities, the military buildings that tourist offices call "Vauban stars" offer straight contours, angles of various openings and walls of variable lengths (but not random), whose imposing size excludes any direct measurement.

This work was conceived within the project *War and Sciences*, which was developed as part of the commemorations of the First World War. Initially focused on the use of sciences in the military context, this project is mainly concerned with physical issues (movement of projectiles, transport of heavy objects...), medical matters (use of deadly gases, invention of radiography, aid to the "broken faces") and even ethical problems (ration of the soldier, strategic use of animals).

The inclusion of a mathematical component was done naturally. As the first aim of this project, focused on the Great War conflicts, was to question the role of scientists in the face of threats from industrial war, we found it judicious to establish a parallel with the situation in Europe at the time of the wars of religion. This comparison is an opportunity to discover geometry as a language that every engineer understood and as a tool of reasoning essential to the affairs of men. Without describing in detail all aspects of the project, we will focus on the time when fortification became geometric, this seventeenth century that saw engineers resort to logical demonstrations inspired by Euclid's *Elements* to establish the validity of their layouts, and this long before Vauban's birth.

Historical Context

Attack and Defence of Places: A Technological Crossover

The dynamic of war originates in the opposition of two poles that mutually oblige each other to constant evolution: attack and defence, the latter being understood as a response to the former. The principles of fortification have varied over the centuries before being gathered into a set of rational precepts at the beginning of the seventeenth century. We could not explain it better than by quoting the first plate (Fig. 9.1) of a Dutch book of this time (Dögen, 1648), whose author traces the history of his discipline through very explicit illustrations.

When human beings left their ancient natural shelters (caves, rocky spurs) of prehistoric times to settle down and create housing communities, they surrounded these with thick walls surrounded by ditches (illustration captioned *Figura* I, top right) reproducing those of the caves: "Art imitating Nature easily found the Ditches & the Ramparts" (Dögen, 1648, p. 8). But these walls had a serious flaw: when the enemy reached their foot, it was impossible for the inhabitants to defend themselves without exposing themselves in the same time. Consequently, the enemy could begin his work of undermining without fearing the defenders. These benefited from the

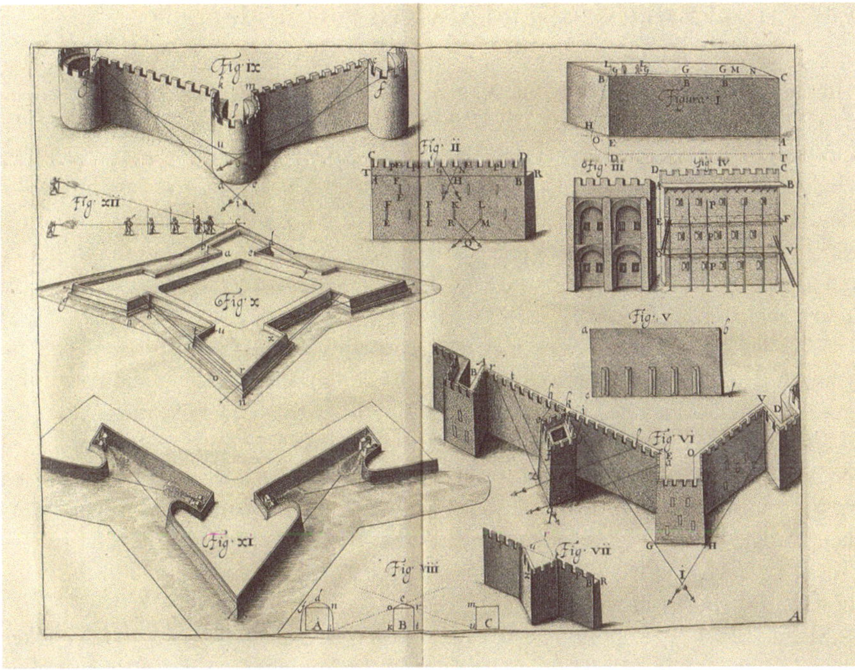

Fig. 9.1 History of fortification by Matthias Dögen (1648), municipal library of Dijon (Photo: Frédéric Métin). Or https://gallica.bnf.fr/ark:/12148/bpt6k15213149?rk=64378;0 ©Gallica

invention of battlements and loopholes which allowed them to act while remaining covered. But, as can be seen very well on the illustration captioned *Fig.* II, there remained spaces impossible to defend due to blind spots. Moreover, the ancient armies had perfected projectile launching devices (catapults, scorpions...), which made them capable of causing enormous damage within fortified enclosures and opening breaches through which invading cities became easy. A way had to be found to keep the aggressors at a safe distance from the walls, making any presence in the ditches perilous. The solution to this problem was the *flanking* of fortified enclosures.

We do not know the historical origin of this principle, but ancient treatises attest to its use at the time of the Roman conquest: it was enough to build towers breaking with the straightness of the wall, stone outgrowths allowing defenders to attack from the rear when the enemies ventured into the ditches. But about square towers of ancient cities (*Fig.* VI) as well as round towers of medieval castles (*Fig.* IX), the question of blind spots remained. No form of tower allowed to completely eliminate the areas in which a brave attacker could find himself sheltered from the defenders' arrows. The invention (or rediscovery) of gunpowder at the end of the Middle Ages in the West would force architects to completely revisit this question.

The "Metal Ball Crisis" and Modern Fortification

Although contested by current historians (Rogers, 2018), the expression "metal ball crisis" is explicit. Medieval stone balls had little chance of hitting a target precisely, and the gun muzzles that propelled them caused damage first to the gunners who handled them. However, it was indeed the improvement of cast iron cannons under the impetus of the Bureau brothers and the decisive use of a significant artillery at the Battle of Castillon in 1453 that allowed King Charles VII of France to put an end to the Hundred Years War. The French artillery would then work wonders in the Italian campaigns.

Whether it is a "crisis" or not, we leave this debate to historians: see for instance (Rogers, 2018). However, it remains certain that the development of this modern artillery changes the very conditions of siege warfare: cannons now fire in a straight line, and metallic projectiles impact is highly unrivaled compared to medieval mortars. The height of towers becomes problematic, as it no longer has the advantage of putting them out of reach of the infantry, formerly vulnerable in their ascent. On the contrary, it makes these proudly erected works ideal targets for distant artillery, and when they collapse, they crash on the inhabitants of the cities they were supposed to protect.

Medieval keeps are no longer relevant. This understanding seems to have taken place in Italy, at the time of the campaigns of the King of France, and then spread throughout Europe, especially in the Netherlands. To adapt to the new conditions of siege warfare, Italian architects invented the *bastions*. These are low towers designed to receive defensive equipment (cannons, balls), always following the principle of flanking, but with a shape designed to eliminate unreachable areas where attackers could place their explosives without fearing the defenders' fire. What is this shape? Along with all his contemporaries, Dögen recommends the pentagonal shape (*Fig.* X and *Fig.* XI of Fig. 9.1) which "leaves nothing unsheltered from the enemy's effort" (Dögen, 1648, p. 13). This Italian invention soon spread, favored by the recruitment of Italian engineers throughout Europe.

The geometric aspect of the bastion profiles is reinforced by a theorisation centered on the creation of bastions for regular polygons. Even if all fortresses do not present this regularity, learning how to build bastions implies being trained through standardized and repetitive exercises on the pentagon, hexagon, etc. The authors of fortification treatises qualify this first approach as "regular fortification", which is then adapted to the actual conditions on the field with "irregular fortification", but this is no longer a question of choosing the appropriate constructions among those that have been described and taught in the "regular" part (Buisseret, 2000; Faucherre, 2008).

The Italian treatises we are aware of do not fully justify the methods of tracing bastions; sometimes, you would even wonder whether the shapes are not generated as the occasional result of numerous traces whose engineers would have chosen the ad hoc lines. This is what one can assume from the drawings produced by some Italian engineers at the end of the seventeenth century, as evidenced by Fig. 9.2,

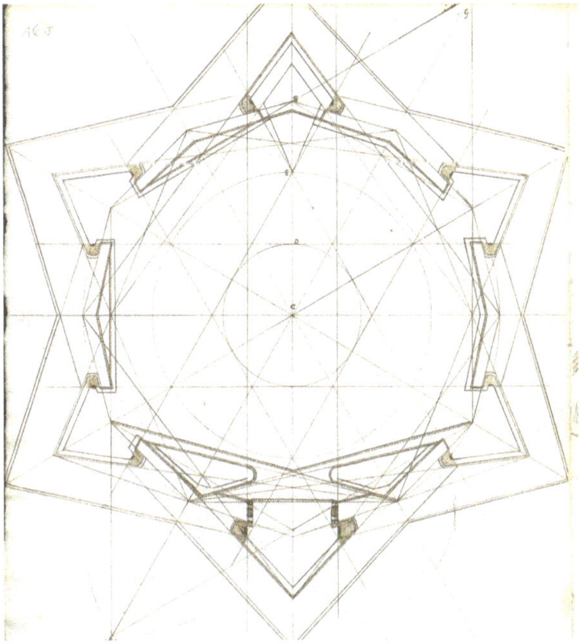

Fig. 9.2 From an anonymous Italian manuscript, end of the seventeenth century, Jagiellonian Library, Krakow (Photo: Frédéric Métin)

extracted from a manuscript collection preserved at the Jagiellonian Library in Krakow.

The lack of details on construction protocols can undoubtedly be partly explained by the strategic necessity of keeping them secret, which will still prevent, a century later, the famous Vauban from publishing his method of fortifying places, as he explains in the dedication to the Duke of Burgundy of his manuscript of *Treatise on Sieges*:

> My Lord, I am trembling as I take the liberty of dedicating this work to you [...] the favour that I dare to ask you, My Lord, is to be willing to take the trouble to read this treatise with attention, and that it pleases you to keep it for yourself, and not to share it with anyone, fearing that some may take copies which could pass to our Enemies and would perhaps be better received than they deserve. (Sébastien Le Prestre, Marquis de Vauban, 1704, pp. 3–4)

On the other hand, military or professional secrecy cannot explain the multiplicity of lines in Fig. 9.2. The construction seems to call for no angle transfers, but only for traces of straight lines connecting the tops of several adjusted pentagons. Like other Italian manuscripts preserved in European libraries, this one does not contain any justification for the traces. These therefore seem to correspond to a prior vision of the desired shape, which should be obtained by choosing the appropriate lines among all those that arise from crossings and intersections. This would therefore be, ahead of time, a restoration of figure in the sense of contemporary didactics (Vendeira & Coutat, 2017; Mithalal & Balacheff, 2019). In the current state of our knowledge, Jean Errard was the first military architect who established his construction principles on an analysis of the forces in presence and deduced his traces from precepts that he himself drew from it (Lallemend & Boinette, 1884).

Jean Errard and the Geometrisation of Fortification

Overshadowed by Vauban, Jean Errard is nevertheless the introducer in France of Italian-style bastioning. At least he is the first who has described the methods in a scientific treatise, *La Fortification réduite en art et démontrée* [Fortification reduced into art and demonstrated], published in 1600 on the command of King Henry IV, for whom Errard was the main military engineer. The *reduction into art* mentioned in the title of the treatise consists precisely in describing exactly the methods of military engineers, both in terms of materials and actual construction, from the project on plans to the effective realisation of the fortress. As for the *demonstration*: this word must be taken both in its geometric meaning, justification using mathematics, and in the sense of picturing the final achievement by staging their military function (Vérin, 1993).

From a meticulous description of contemporary artillery followed by comparing the forces at hand (the firepower of the aggressor, the human forces of the defenders), Errard draws four geometric rules that he calls the *maxims* of fortification, fully justified. Let's illustrate, for example, the most important maxim, according to which the angle of the bastions must be right (Fig. 9.3). As the author explains, the batteries of enemy cannons are always pointed at the wall perpendicularly, as this is the position in which the cannonballs do the most damage. If the angle of the bastion is acute (Fig. 9.3, on the right), the tip does not contain enough earth and the angle is easily "ruinable" by the cannons, while with a right angle (Fig. 9.3, on the left), the amount of earth behind the wall is always the same. It is then sufficient for the side of the bastion to be long enough for this amount to withstand the bombardments.

But how then were the bastions built? Let's follow the procedure given for the hexagon by Alexis Errard (Jean's nephew) in his edition of the *Fortification*, proposing the following diagram (Fig. 9.4) to facilitate reading.

> Let it be proposed to fortify a Hexagon, since the Hexagon is divided into six equilateral triangles. Let the equilateral triangle ABC be described on AB, then let the angle CAD be made of forty-five degrees: Let the line AE be made equal to the line BD, then let BE be

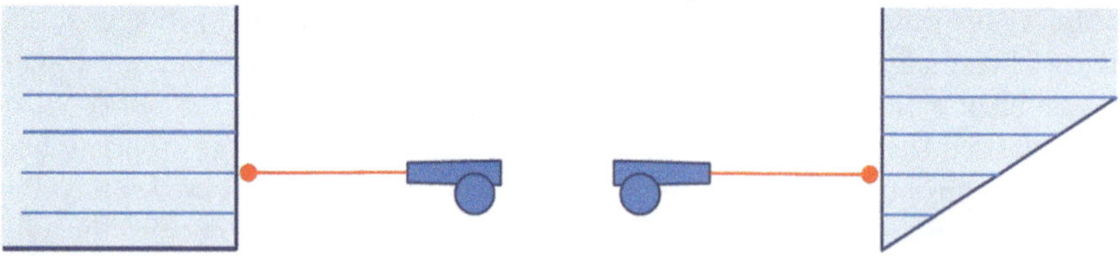

Fig. 9.3 Justification for the choice of the right angle at the tip of the bastions

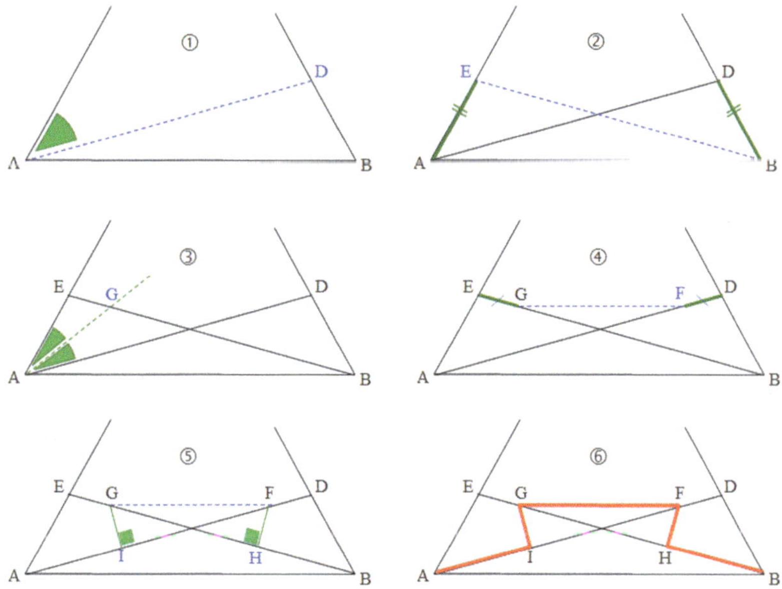

Fig. 9.4 The construction protocol of Jean Errard in six steps

> drawn. Let the Angle EAD be divided equally by the line AG, & let DF be taken equal to EG, & let the Curtain GF be drawn: as also FH perpendicular to the line BE. Let AI be taken equal to BH, & let the line GI be drawn perpendicularly like FH. Thus will be described the two half Bastions AIG, & FHB. (Errard & Errard, 1620, p. 24)

Thus, the line $AIGFHB$ is the profile to repeat five times to obtain the hexagonal fortress of Fig. 9.5 (approximations being put aside).

In our interpretation (Fig. 9.4), we have chosen one of the two possible procedures at step 5, constructing I in the same way as H, while Errard rather proposes to carry the length BH from A to I. Let us note here that at no point did we have to measure any length whatsoever. The compass is only used to *transfer* them, which is its primary function as a drawing instrument.

In the rest of his text, Jean Errard will demonstrate the adequacy of the form obtained to the specifications that constitute his maxims, using exclusively the universally accepted theorems of Euclidean geometry. For example, a repeated application of the Pythagorean theorem will allow him to precisely evaluate line AF, named *line of defence*, and thus prove that it does not exceed the range of the musket (estimated at 120 toises), in accordance with the requirements of the second maxim. A few years later, Samuel Marolois and the Dutch fortifiers, from whom Vauban will draw his main ideas, will take advantage of the efficiency of trigonometry to extend Errard's work—see for example (Marolois, 1615) and (Dögen, 1648). But this goes far beyond the scope of primary school in which we have experimented with the use of these texts, as we now relate.

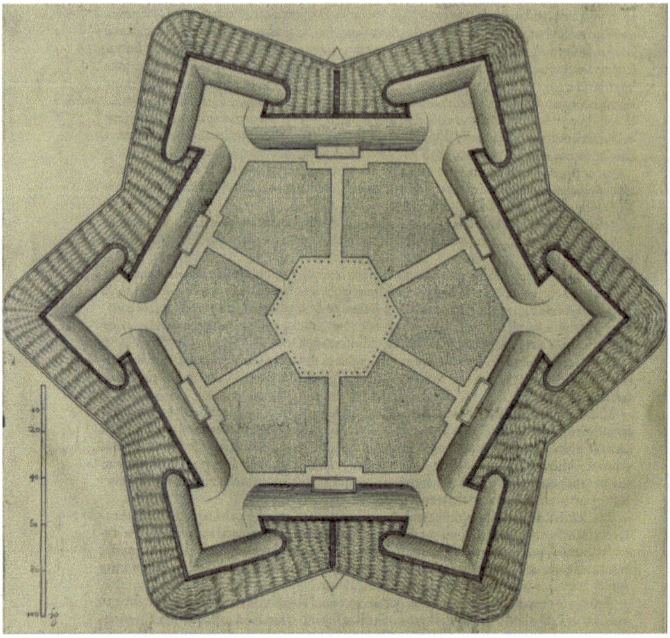

Fig. 9.5 Jean Errard's fortified hexagon (Errard, 1600, p. 30), Jean Jeukens Library, Bar-le-Duc (Photo: Frédéric Métin). Or: https://gallica.bnf.fr/ark:/12148/bpt6k1512241w/f12.item.r=Errard,%20Jean ©Gallica

Scientific and Pedagogical Aspects

The Project *War and Sciences*

Focused particularly on the First World War, the project has subtitle "When sciences mobilise" and mainly aims to allow primary school pupils to see through science the essential elements of the First World War, and consequently to better understand their own role as citizens. The scientific concepts addressed fall within the field of technical objects, in a first approach to the principles of lifting, flight, propulsion and communication. More generally, pupils are invited to look at the major scientific advances linked to the war and the involvement of renowned scientists in the nations in conflict (such as Marie Curie, Paul Langevin or Fritz Haber in particular). The activities are based on observation, the formulation of hypotheses, experimentation and presentation of their results, which corresponds to the investigative approach, so rarely chosen for mathematics activities. The teacher summarises the essence of her pedagogical objectives in the following diagram (Fig. 9.6).

This open guidance in an investigative approach is quite rare in mathematics activities. Moreover, the geometry workshop leaves little room for experimentation, even if it leaves a good part to the initiative of the pupils. However, in the context

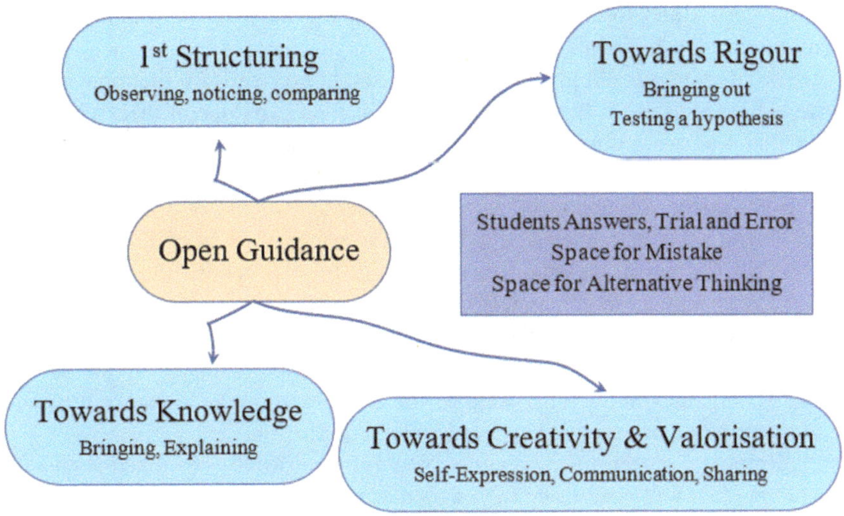

Fig. 9.6 The teacher's posture towards the class

of workshops dedicated to science and technology, the practice of geometry takes on a concrete meaning when it comes to thinking about the effective construction of fortresses, with the following problem: how to build a wall resistant to the projectiles that other groups will have learned to launch?

The thematic day *attack and defence* was preceded by a general presentation, focused on a timeline (Appendix 1) that the pupils were invited to fill in collectively, first by small groups then by bringing together the two final classes of the school (pupils aged 10 and 11) for a collective debate. The preparatory work consisted in thinking about the materials usable for the rampart they were going to create and test against the enemy artillery. Thus, the geometric construction of a hexagonal fortress according to Jean Errard is not conceived as an activity isolated from any context. On the contrary, it allows to show the usefulness of mathematics both in the modelling of field problems and in their treatment.

The Stakes in Geometry

The framework of the project offers a contextualisation and a problematisation of the drawing of a geometric figure that a simple mathematics exercise would not have allowed (for a general discussion, see Ransom (2012, 2015)). The shape of the seventeenth century fortified enclosure has the advantage of existing in reality, even if this reality only persists through the ruins of monuments. However, satellite images combined with reproductions of period plans provide a striking insight into the star-shaped forms and their current management in the urban landscape (demolitions, developments, various reuses).

We are therefore explicitly situated in a geometry of maps (in the topographical sense) which legitimises the study of rectilinear figures by one of their reasons for being, that is to say as constituent elements of mass plans of buildings or only a part of these buildings (see, for example, the United States Department of Defense, the famous Pentagon). We could just as well consider a geometry of plans on a smaller scale, in which the rectangles would be tables to arrange in a room also rectangular.

Once the real context is established, the work focuses mainly on the reconstruction of a form subject to constraints on the sizes and angles (the only determinants of rectilinear plane figures), respecting in some cases conditions of symmetry (Duval, 2015). This reconstruction is conditioned on the one hand by the analysis of the existing, and on the other hand by the available drawing tools, the graduated instruments being not useful (Duval & Godin, 2005).

Pupils must discover that a polygonal enclosure is partially bastioned from a first broken line that will need to be reproduced as many times as there are triangles in the considered polygon. This first (visual) step of analysing aerial images or Ancien Régime military maps can rely on the rich iconographic resources put online in recent years (bibliography and webography).

It is then a matter of embarking on the construction in two stages: the geometric drawing of the basic element, then the reproduction of this as many times as necessary, and in positions to determine. This determination is the most delicate part of the activity. Indeed, although the decomposition of the polygon into triangles is the subject of a fruitful autonomous research phase, the reverse process of reconstructing the complete form from the first drawing is less natural. Pupils especially need help when it comes to finding precisely in which position and where to place this basic element (that is to say the red profile of Fig. 9.4), taken in the form of a template. Unless they choose to repeat the tracing program itself, positioning each of the triangles composing the polygon as they did with the first one, or even to draw the outline freehand between two consecutive vertices of the polygon, visually reproducing what they initially obtained.

Whatever the case, the question of the drawing gesture is crucial, and consequently that of the instruments to use. If we have chosen to rely on the text of Jean Errard, it is because it does not appeal to any quantitative measure, except the 45° angle which disappears from our version (Appendix 3). The form is indeed obtained only by dividing angles into two equal parts (from the right angle to its half, then its quarter), and by reproducing lengths. As for the angles, we can leave the initiative to the pupils, who will know, from a right angle, how to obtain its half by symmetrical folding, then half of this half. For distances, it is possible to cut strips of paper whenever it is necessary, since it is only a matter of reproducing certain lengths without any of them to be used to measure the other ones. The use of the compass in this context is also justified as it corresponds to the transition recommended in the programs between the transfer of lengths on an already drawn line and the understanding of the notion of a circle.

We hope this work allows an awareness of the reiteration of a tracing process, in the manner of a computer procedure repeated in a loop of an algorithm. The economy of means that iteration allows, especially in the geometric field, can be the basis for further explorations using software like Scratch, explorations that we have not yet been able to implement, the use of computing in this context is still in its early stages in the terminal classes of primary school.

One of the research paths relies on the fortification method of Blaise-François Pagan, Vauban's mentor, of which we have described otherwise (Guyot & Métin, 2006) the use of a single broken line valid in the case of all polygons, from the hexagon to the dodecagon.

Implementation in the Classroom

Loads and Projectiles

Two workshops of the project *War and Sciences* were consistent with a mathematical approach to attack and defence, so they were conducted in parallel with the geometry and construction workshops, for a final confrontation of the groups of pupils at the foot of the rampart built in the physical education room.

The first science workshop is about the lifting of loads. Starting with observing photographs showing soldiers or civilians during First World War handling shells or heavy loads, the pupils identify several lifting techniques: the inclined plane, the pulley, the lever. Then, with elastic bands and different weights to lift, the pupils are led to observe that the heavier the weight lifted, the longer the elastic stretches. We can therefore roughly measure the effort made to lift the different loads and compare the data obtained. Then, the pupils must vary the slope of the ramp, the rotation point of the pivot or the sequences of pulleys to observe the effects on the length of the elastic and interpret their measurements.

The second science workshop focuses on the construction of devices used to launch various projectiles (marbles, bullets, balls...). A preliminary musical introduction highlights the importance of artillery in the conflicts of the First World War. The teachers mention the *Big Bertha* and the *Parisian Cannons* and they stimulate pupils into questioning the account of French aviator René Fonck about Paris bombing on 23 March 1918: the French were trying to flush out the high-altitude bomber planes, as they could not believe in the possibility of shots from 120 km away. After a general questioning on the range of the cannon, the pupils experiment with various means of propelling the projectiles: balloon cannons, vortex cannons, elastic cannons, made with the means at hand, cardboard, cylindrical boxes, etc. The use of video in slow motion allows questioning the pupils on the trajectory of projectiles, which gives results of remarkable relevance in their productions (Fig. 9.7).

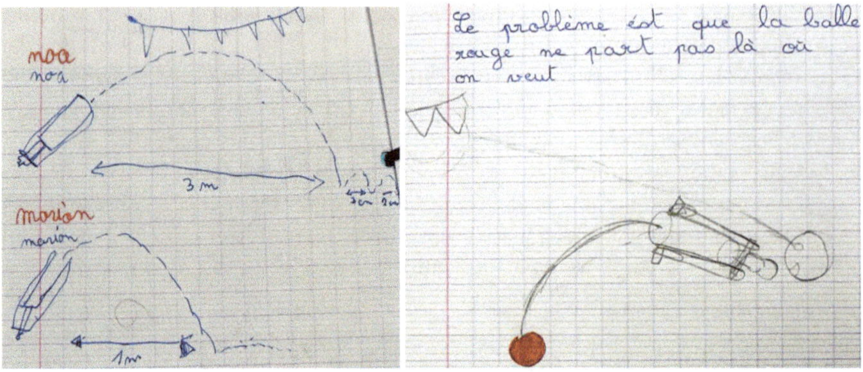

Fig. 9.7 The trajectory of projectiles [on the right (English translation): "The problem is that the red ball doesn't go where we want it to go"] (Photos: Frédéric Métin)

The Geometry Workshop

The actual construction workshop is preceded by a presentation of regular polygons and polygonal fortresses of the seventeenth century, built on the plans of engineers of various origins (Italian, Dutch, French) and an increasing number of sides. The pupils examine satellite images of the citadels of Spandau (square, Berlin), Bourtange (pentagon, Netherlands), Lille (pentagon, France), Neuf-Brisach (octagon, Alsace) and Palma Nova (enneagon, Italy). They understand that the regularity of the shapes was favoured with the aim of presenting no flank weaker than the others, but the choice of pentagonal bastions remains mysterious to them. It must be said that these choices are mysterious for us as well, as author from the seventeenth century never justified them in reality. Even Jean Errard seemed to feel his choices went without saying. However, it is possible to verify that the bastions of Errard's fortresses are in fact truncated squares, by reconstructing from the two half-bastions of Fig. 9.4 the complete circuit of two consecutive bastions (Fig. 9.8).

When it comes to undertaking the layout following Errard's construction program, the original text presents several difficulties, starting with the expression and spelling of Middle French. This obstacle is easily overcome by a modern rewrite of the excerpt (Appendix 2), taking into account the fact that the measure of the 45° angle isn't in the curriculum: so here it will be a "half-right", a neologism that could very well have been coined in Errard's time. For reasons of time, we chose the prior rewrite, but this work could have been the subject of an activity with the pupils. Similarly, the initial equilateral triangle is given (Appendix 3), although its construction is achievable by the pupils themselves.

The second difficulty still lies in the omnipresence of angles, quantities that pupils are still not used to manipulating, especially since it is a matter of drawing the bisector of angle CAD, which doesn't belong to the curriculum. As indicated above, we have chosen to provide the pupils with two templates representing the

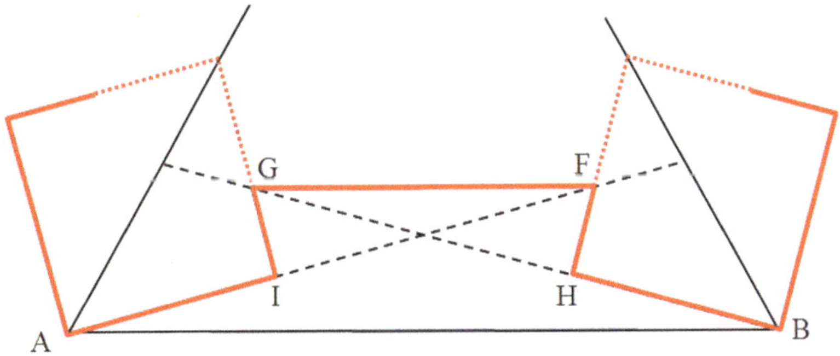

Fig. 9.8 The implicit shape of Errard's square bastions

Fig. 9.9 First step of the geometric construction

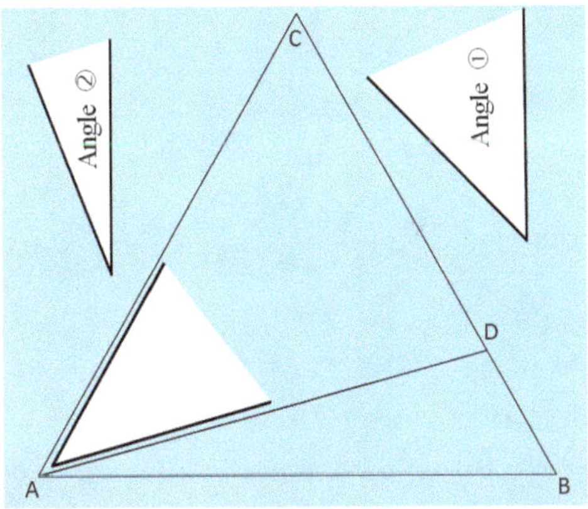

two angular sectors they would need. During the activity, some pupils were able to use juxtaposition to verify that angle 1 represented twice the angle 2, in other words, by using angle 2, they would indeed divide CAD into two equal parts.

The question of how to transfer distances is also a subject of debate. Errard is not very explicit at this point, only asking to "draw line AE equal to BD". Where should E be located? The argument of regularity that was put forward at the beginning of the workshop suggests a symmetry of constructions, which instructs to put E on $[AC]$ and to complete the figure afterwards according to the same principle.

As one can imagine, once the task is understood the pupils want to use their graduated ruler rather than the compass. This standard use of the compass is not very common in school, as if the lengths to be transferred were necessarily measurable quantities. Conscious of the possible difficulties of children in the use of instruments, we prepared a slideshow reviewing all the steps to be taken, from which Figs. 9.9, 9.10, 9.11, and 9.12 are extracted. They are to be compared with Fig. 9.4, which focused only on geometric objects, without reference to construction instruments.

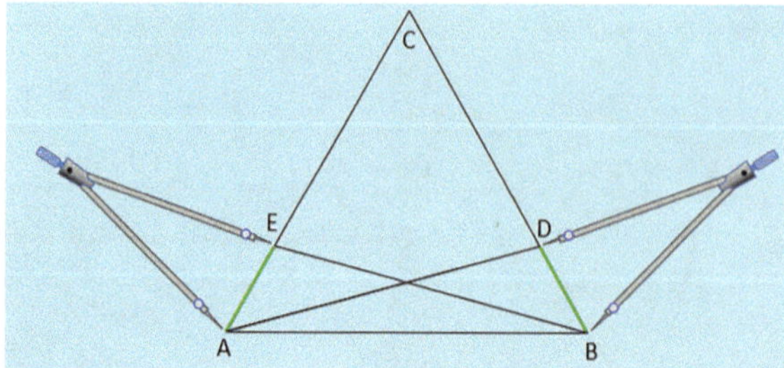

Fig. 9.10 Second step of the geometric construction

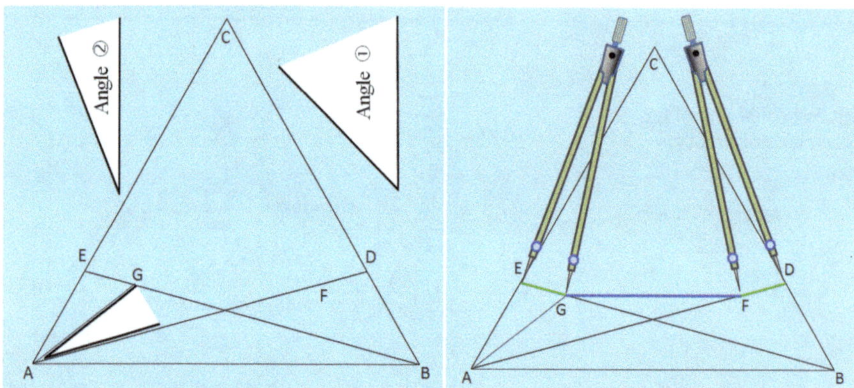

Fig. 9.11 The tracing of the curtain wall $[GF]$

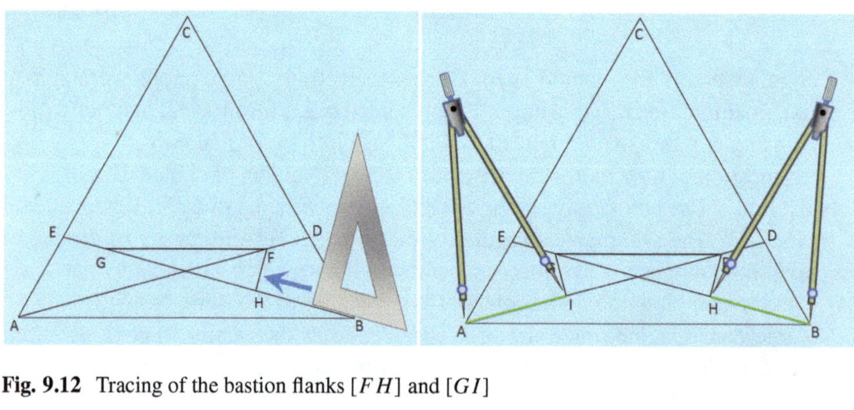

Fig. 9.12 Tracing of the bastion flanks $[FH]$ and $[GI]$

Once line $AIGFHB$ is obtained, all that remains is to reproduce it to obtain the complete outline of the fortified Errardian hexagon. Easier said than done... Here the pupils are left free to their procedures for the complete construction which will be finalised outside of class time. Some will choose to make a template of the constructed line, others will reproduce the tracing algorithm five times, some will only redo the freehand drawing. In order not to limit the exercise to simple drawing, the completed enclosure leads to an imagination activity: how to arrange the fortress to withstand a siege? In other words, how to create a city within the ramparts?

Ways to Go Further

Designing a Fortress

It was a city... or a garden, depending on the request of some pupils who were not inspired by urban planning. A paper support with a dotted hexagon had been prepared for those who needed it (Appendix 5). Examination of the productions led us to notice that many pupils had chosen to use a template. This strategy is highlighted by the irregularity of the bastion sides reproduced from peak to peak (Fig. 9.13).

But whether the bastions are in accordance with the Errardian construction or not, the differences within the same drawing are minimal (Fig. 9.13). A remarkable aspect of productions A and B is that they take very little account of the shape of the fortress to draw its interior: for pupil A, the interior is octagonal, like the fortress of Neuf-Brisach, and for B, it is a rectangle, containing its own castle for enhanced security. In both cases, the interior figure is independent of the shape of the enclosure, notably not having the same centre.

The centre of the figure is however essential in the other productions which are organised around it. The radial aspect of the fortress is perfectly integrated in the following four productions (Figs. 9.14 and 9.15), even if the central place of citadel E is not exactly in the centre of the figure. Note for C a perfect distribution of cannons on the curtains for defence from all sides, while E has favoured a local layout with four houses, like in a real citadel of limited size. On the contrary, the

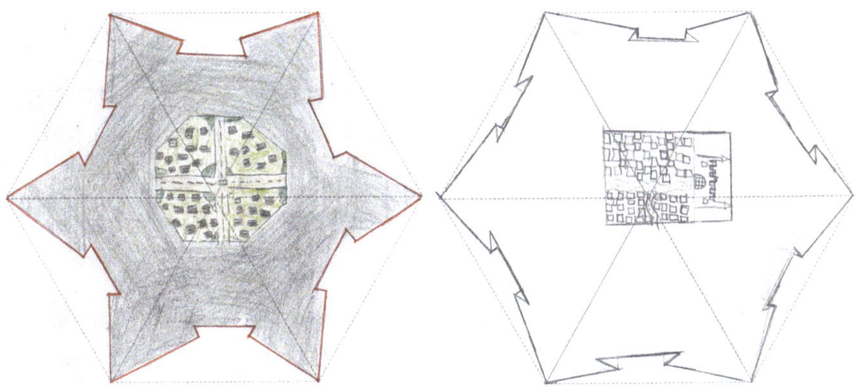

Fig. 9.13 Productions A and B (Photos: Frédéric Métin)

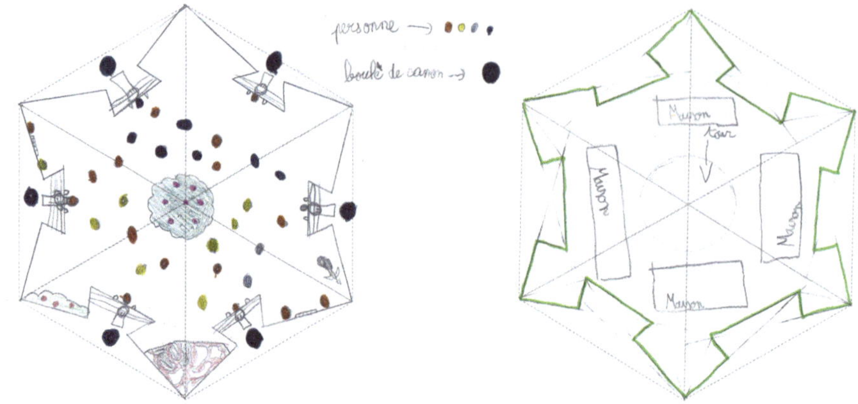

Fig. 9.14 Productions C and D [on the left: "persons" and "cannonballs"; on the right "house" (rectangles) and "tower" (circle)] (Photos: Frédéric Métin)

Fig. 9.15 Productions E and F [top to bottom: "church"; "house"; "building"; "fountain"; "pool"; "shop"; "bench"; "street"; "hospital"; "tree"] (Photos: Frédéric Métin)

superproduction E considers all aspects of large-scale planning, with residential areas, building zones, forest, fountains, and even a church, in the centre as it should be.

Admittedly, a sort of forest was already in the centre of citadel C, but this one only contained soldiers and cannons. Finally, note for E and F the existence of streets starting from the central square and leading to the ramparts, which corresponds exactly to the radial plan of Errard's fortresses in terms of the main arteries.

Restoring a Citadel

The plans of fortresses generally presented to the public elicit few geometric comments, as they are only images illustrating the remarks. The name—stars—given to citadels and the "geometric" qualifier often attached to them rarely lead to truly geometric considerations, in the mathematical sense of the term. As a rule, archaeologists and admirers of Vauban are content with underlining the parallelism of successive enclosures, and geometry is only invoked to highlight the angular aspect of the contours, or even the qualities of the architect as a geometer (Griffith & Dennis, 2006).

However, as Duval points out, one of the issues raised by teaching geometry is how to develop a view centered on the properties of figures rather than centered on their shape (Duval, 2015). One way to promote the acquisition of this perspective is to reconstruct given figures with a constraint on the instruments used, preferably incomplete templates and stencils.

Furthermore, studies on figure restoration as (Vendeira & Coutat, 2017) have shown the full interest of reconstructing incomplete figures for the acquisition of geometric concepts, through the implementation of strategies using these concepts. Why not a figure restoration complementing a military monument restoration?

Indeed, Errard's masterpiece offers this profile of an incomplete fortress. It is the pentagonal citadel of Amiens, of which he gave the plans on the order of Henry IV following the siege and the reconquest of the city from the Spanish in 1597. Various contemporary plans show that the project corresponds exactly to the bastioned pentagon depicted in the *Fortification* of 1600, even if the shape was not exactly respected for the bastions and curtains of the eastern side, built after Errard's death in 1610. As this part was destroyed in the 1950s for the piercing of an avenue, the citadel now presents an incomplete form which would be perfect for a reconstruction activity.

The activity, not yet tested, of restoring the figure of the citadel could be conducted with pupils already familiar with Errard's hexagon bastioning. In this perspective, they would be given complete freedom to reinvest their knowledge of the construction procedure of the bastions, a procedure to be repeated twice to obtain the complete plan of the citadel given by Errard in 1597.

Conclusion

The shape of Italian-style bastioned citadels allows for working on unusual geometric figures, examples of which can be found in the military history of Europe. The plans and satellite views offered on *OpenStreetMap* or *Google Earth* are valuable educational tools for encouraging questioning and inserting geometrical exercises into a strategic problem full of historical and geographical meaning. The strong evocative power of images is an asset here for observing figures that escape a strictly confined existence in the mathematical domain. It's an additional opportunity to highlight the usefulness of mathematics in various everyday questions, even those that involve the survival of states.

The inclusion of a geometry workshop in one day of the project *War and Sciences* was precisely the opportunity to show how geometry could have allowed the transcription of defence problems related to the progress of artillery and to seek the best adapted solutions, finally to demonstrate their adequacy to the constraints of the terrain. This situation, even if it no longer corresponds to current practices, is likely to reinforce the feeling the usefulness of geometry for our pupils and to make them feel the deep links between mathematics and other scientific disciplines.

The difficulty of using Errard's text comes from his systematic use of constructions based on angles, which concepts are not very present in primary school programs. The use of the protractor, which allows to broaden the field of angles used, only comes into play in secondary school, as do activities of angle reproduction. However, pupils can appropriate notions usually approached through measurements well before mastering certain measurement tools in geometry, as the progressivity markers of the program allow. Isn't it indeed a less restrictive approach than evoking the equality by superposition of angular sectors even before of the equality of angle measurements? That this superposition may be impossible can be seen as an opportunity to engage pupils in seeking other means to establish equality, starting with the transport of templates and then by evaluating the opening of the angle.

Our work needs to be extended to secondary school classes, as the difficulties related to angles have not allowed us to consider a drawing using software like Scratch, which would be a natural continuation of our activity. Indeed, if the pupils have understood the challenge of decomposing the contour of the ramparts into a basic element reproduced several times, the time and resource savings thus achieved would have been much more noticeable in the context of a block of instructions reusable at will. This first approach to iteration would also have allowed us to realise the need to anticipate the number of repetitions of the initial protocol, in our case: five repetitions for the hexagon, but more (and in a number to be determined, with possible generalisation) in the case of the following polygons.

A vast project remains open.

Acknowledgments This chapter was written by Frédéric Métin, with the help of two teacher trainers, Agnès Golay and Agnès Gateau (all three from the IREM of Dijon, France). The project War and Sciences was developed by Agnès Golay at the Resource Centre for Sciences of Côte d'Or (CDRS 21) and the experiments were carried out at the Petit-Bernard primary school in Dijon with Isabelle Arbez and Sandrine Gillet, teachers at this school and Jean-Luc Pernette, mathematics/science teacher at the Bonaparte high school in Autun and head of the "Cycle 3/4" group at the IREM of Dijon.

Appendix 1: Concepts and Skills Addressed in This Chapter

Concepts Covered

- Recognise, name, describe, reproduce, represent, construct some geometric solids and figures
  - Recognise, name, compare, verify, describe simple or complex figures (arrangement of single figures)
  - Recognise and use some geometric relationships: parallelism, alignment and belonging; equality of lengths
  - Carry out, complete and write a construction program
  - Plane figures and solids, first characterisations: triangles, including special triangles (right-angled, isosceles, equilateral)
- Compare, estimate, measure geometric quantities with whole numbers and decimal numbers: length (perimeter), area, volume, angle
  - Identify angles in a geometric figure
  - Compare angles
  - Reproduce a given angle using a template
  - Draw a segment of a given length or transfer the length of a segment
  - Complete a figure by axial symmetry

Skills Developed

- **Search and research skills**: Extract and organise the useful data for problem solving from various sources: texts, tables, diagrams, graphs, drawings, schemes, etc.
- **Reasoning**: Analyse a plane figure from different aspects (surface, contour, lines and points).
- **Mathematical representation ability**: Use tools to represent a problem: drawings, schemes. Reproduce (to scale or not) a figure from a model and already drawn elements. Situations of reproduction or construction involving basic measuring and drawing gestures and knowledge about usual figures.
- **Modeling**: Recognise real situations that can be modelled by geometric relationships (alignment, parallelism, perpendicularity, symmetry).

Appendix 2: Basic Figure Distributed to Pupils (A5 Format)

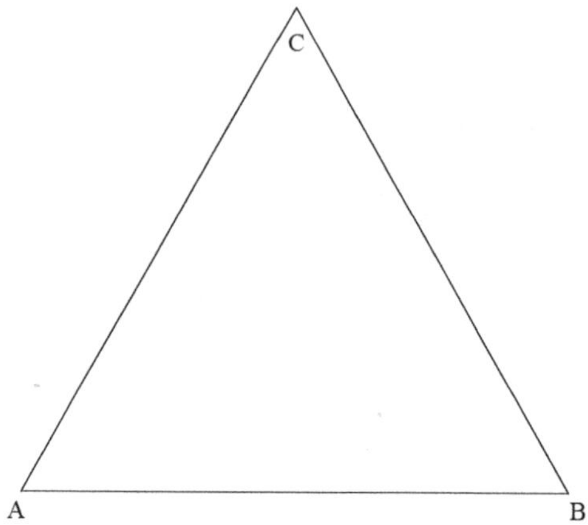

Appendix 3: Adapted Text by Jean Errard Distributed to Pupils

To Fortify the Hexagon, According to Jean Errard de Bar-le-Duc (1554–1610)

Let it be proposed to fortify a hexagon, as the hexagon is divided into six equilateral triangles. Let the equilateral triangle ABC be described on AB, then let the right angle CAD be made.

Let the line AE be made equal to the line BD, and then let BE be drawn.

Let the Angle EAD be divided into two equal angles by the line AG, let DF be taken equal to EG, and let the curtain GF be drawn.

Let FH also be drawn perpendicular to the line BE, and let the line GI be drawn perpendicularly like FH.

Let AI be taken equal to BH.

Thus the two half-bastions AIG & FHB will be described.

Appendix 4: Templates Distributed to Pupils

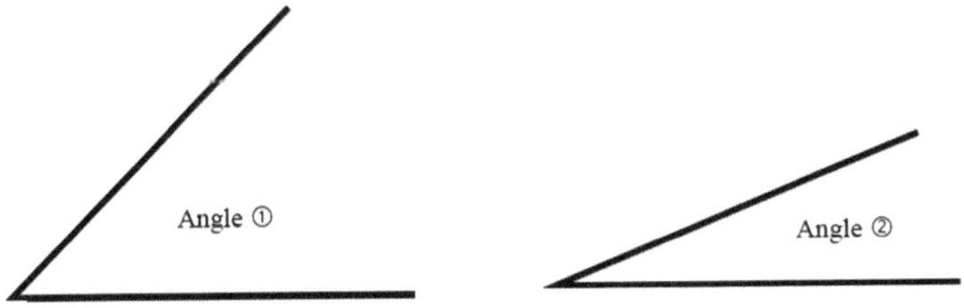

Appendix 5: Support for the Construction of the Enclosure to Be Urbanised

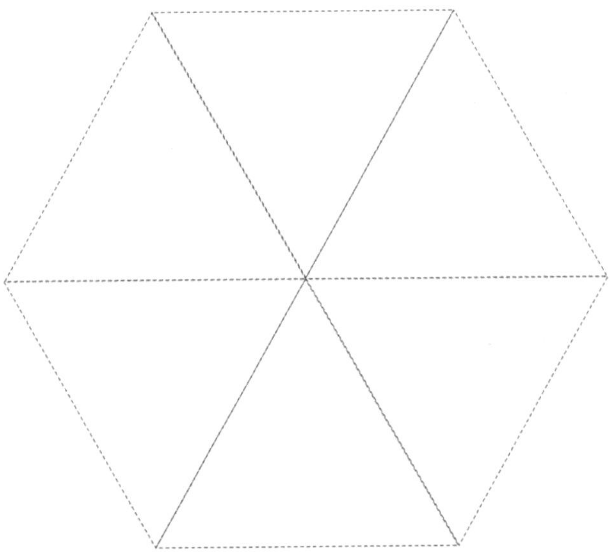

References

Buisseret, D. (2000). *Ingénieurs et fortifications avant Vauban: l'organisation d'un service royal aux XVIe-XVIIe siècles*. Comité des travaux historiques et scientifiques.
Dögen, M. (1648). *L'Architecture militaire moderne ou Fortification...* Louys Elzevier.
Duval, R. (2015). Geometrical pictures: Kinds of representation and specific processing. In R. Sutherland & J. Mason (Eds.), *Exploiting mental imagery with computers in mathematics education* (pp. 142–157). Springer.

Duval, R., & Godin, M. (2005). Les changements de regard nécessaires sur les figures. *Grand N*, *65*, 7–27.

Errard, J. (1600). *La fortification reduicte en art et demonstree...* [at the author's].

Errard, J., & Errard, A. (1620). *La fortification demonstree et reduicte en art...* [Pierre Delon printer].

Faucherre, N. (2008). *Places fortes, bastion du pouvoir*. REMPART–Desclée de Brouwer.

Griffith, P., & Dennis, P. (2006). *The Vauban fortifications of France*. Osprey Publishing.

Guyot, P., & Métin, F. (2006). Using history of maths and physics in teachers training: The officer's sciences in the ancien régime. In F. Furinghetti, S. Kaijser, & C. Tzanakis (Eds.), *Proceedings of HPM2004 and ESU 4* (pp. 323–333). Uppsala University.

Lallemend, M., & Boinette, A. (1884). *Jean Errard de Bar-le-Duc, "premier ingénieur du très Chrétien Roy de France et de Navarre Henry IV", sa vie, ses œuvres, sa fortification (lettres inédites de Henry IV et de Sully)*. Thorin & Dumoulin, Comte-Jacquet & Boinette.

Marolois, S. (1615). *Fortification ou architecture militaire tant offensive que deffensive...* Hendrik Hondius.

Mithalal, J., & Balacheff, N. (2019). The instrumental deconstruction as a link between drawing and geometrical figure. *Educational Studies in Mathematics*, *100*, 161–176.

Ransom, P. (2012). Fortifying France: The cities of Vauban. In T. Kerry (Ed.), *Proceedings of "History and Pedagogy of Mathematics 2012"* (pp. 323–333). KSME. http://www.hpm2012.org/Proceeding/Workshop/W2-4.pdf

Ransom, P. (2015). Mainly mathematics: Cross-curricular activities in Key Stage 2. In T. Kerry (Ed.), *Cross-curricular teaching in the primary school* (pp. 83–98). Routledge.

Rogers, C. J. (2018). *The military revolution debate. Readings on the military transformation of early modern Europe*. Routledge.

Sébastien Le Prestre, Marquis de Vauban. (1704). *Traité des sièges, première partie: de l'attaque et la defense des places*. Manuscript 275, Municipal Library of Dijon.

Vendeira, C., & Coutat, S. (2017). Shapes recognition in early school: How to develop the dimensional deconstruction? In T. Dooley & G. Gueudet (Eds.), *Proceedings of the Tenth Congress of the European Society for Research in Mathematics Education* (pp. 677–684). Dublin City University.

Vérin, H. (1993). *La gloire des ingénieurs. L'intelligence technique du XVIe au XVIIIe siècle*. Albin Michel.

Frédéric Métin has taught mathematics in French and in English for 30 years, using historical texts in the classrooms, before joining the Burgundy Institute of Education and earning his PhD in 2016 at Nantes University. The subject of his doctoral dissertation was geometrical fortifications in sixteenth and seventeenth century Europe. Now, his main research subjects are practical mathematics from the late Middle Ages to the Renaissance, including counting with tokens, land surveying, military and civil architecture, etc. For educational and training purposes he is the present leader of the Burgundy "Cormecouli project" which aim is to proposes to primary and middle school students mathematical activities based on the use of original local documents, as medieval accounts of old burgs or properties maps used by wealthy families to increase their grasp on land, as well as notaries archives documenting the average life of more ordinary people.